WORLD BANK TECHNICAL PAPER NO. 394

# Incentives for Joint Forest Management in India

*Analytical Methods and Case Studies*

*I. Hill*
*D. Shields*

*The World Bank*
*Washington, D.C.*

Technical Papers are published to communicate the results of the Bank's work to the development community with the least possible delay. The typescript of this paper therefore has not been prepared in accordance with the procedures appropriate to formal printed texts, and the World Bank accepts no responsibility for errors. Some sources cited in this paper may be informal documents that are not readily available.

ISSN: 0253-7494

I. Hill is a principal agriculturalist in the Rural Development Sector Unit in the South Asia Region of the World Bank. D. Shields is a consultant economist.

**Library of Congress Cataloging-in-Publication Data**

Hill, Ian R.
    Incentives for Joint Forest Management : analytical methods and
case studies / I. Hill, D. Shields.
       p.    cm. — (World Bank technical paper ; no. 394)
    ISBN 0-8213-4143-X
    1. Forest management—India—Case studies.   2. Forest management—
India—Citizen participation—Case studies.   3. Forest policy—
India—Case studies.   4. Joint Forest Management (India)—Case
studies.   I. Shields, D. (Dermot), 1954–  .  II. Title.
III. Series.
SD223.H5    1997
338.1'349'0954—dc21

                                      97-38926
                                      CIP

# Contents

# FOREWORD

Joint Forest Management (JFM) has emerged as an important intervention in the management of India's forest resources. The social and institutional arrangements of JFM have been described in some detail in a number of publications. There is not much work on the quantitative analysis of the financial and economic returns to JFM. To this end, the present study is a pioneering effort and I wish to compliment the authors.

The report prepared by the World Bank in consultation with the Ministry of Environment and Forests, sets out an analytical method to examine the costs and benefits of JFM arrangements. The method has been applied in two case studies on a pilot scale and brings out interesting facts regarding the incentives for various groups of users participating in JFM. However, the findings are based on limited data and further work is required on a wider mosaic of diverse ecological and social conditions to validate the method.

C. P Oberai
Inspector General of Forests
Ministry of Environment and Forests
Government of India

# ABSTRACT

The study examines the economic and financial incentives for various groups of forest users in India, to participate in Joint Forest Management (JFM) arrangements, that is the management of forest resources by government forest departments and local communities. An analytical method is developed and applied to two case studies of communities managing a mixed teak forest system and a sal coppice forest system. The analytical results show that: (i) Economic returns to JFM are considerable in both forest systems; (ii) There is an increase in revenues from the forest to the communities but a theoretical loss to forest departments; (iii) Income flows into the communities increase significantly, though in the sal coppice system this is partially offset by declining income from collection of non-timber forest products (NTFPs); (iv) The benefits of JFM are not always distributed equally, which may result in collectors of firewood and some NTFPs losing, even though overall gains are sufficient to compensate losers. The realization of benefits is dependent on an enabling environment consisting of complex institutional and social conditions, in particular the representativeness and functioning of the village forest committee, the regulatory framework and sharing arrangements, and the regional economic and marketing context. The method needs to be more widely tested in a variety of social and environmental conditions and the results from these two case studies can only be extrapolated with caution. Nevertheless, they point to significant economic and financial benefits to communities, and the need for specific measures to safeguard the interests of those who may lose as a result of unequal distribution of the benefits.

# CURRENCY EQUIVALENTS

Currency Unit = Rupees (Rs.)

US$1.00 = Rs.32.3

# WEIGHTS AND MEASURES

The metric system is used throughout this report

# ABBREVIATIONS

| | |
|---|---|
| CAI | Current Annual Increment |
| CC | Canopy Cover |
| CIP | Current Increment Percent |
| FD | Forest Department |
| FDC | Forest Development Corporation |
| FPC | Forest Protection Committee |
| FWP | Forest Working Plan |
| GOI | Government of India |
| GS | Growing Stock |
| JFM | Joint Forest Management |
| LAMPS | Large-Sized Multi-Purpose Cooperatives |
| MAI | Mean Annual Increment |
| MOEF | Ministry of Environment and Forests |
| NTFP | Non-Timber Forest Products |

Note: Data presented in tables based on field observations or authors' calculations unless otherwise noted

# EXECUTIVE SUMMARY

## Introduction

In India, the term Joint Forest Management (JFM) is used to describe the management of forest resources by government Forest Departments with the participation of local communities. JFM can be defined as follows:

- *Joint Forest Management* is the sharing of products, responsibilities, control, and decision making authority over forest lands, between forest departments and local user groups, based on a formal agreement. The primary purpose of JFM is to give users a stake in forest benefits and a role in planning and management for the sustainable improvement of forest conditions and productivity. A second goal is to support an equitable distribution of forest products.

There is an extensive literature on JFM in India and elsewhere in which social and institutional arrangements are discussed in detail. Technical aspects of forest management has not been systematically examined, whilst quantitative analysis of the financial and economic returns to JFM is largely lacking. There is, therefore, an urgent need to validate JFM models in a variety of social and ecological conditions, so as to provide a firm analytical basis for preparation of future forestry projects.

## Objectives

The main objective of this study is to develop a better understanding of the incentives for communities to participate in JFM, in particular, the economic and financial incentives for participation of various groups of forest users, the stakeholders in the forest resources, both within and outside the community. In response to the National Forest Management Policy, 1988, JFM arrangements have been introduced in a wide range of ecological and social conditions in India. However, for the purposes of this study, it was decided to undertake detailed analysis in a limited range of conditions, rather than more general analysis over a wider range of conditions. The study has, therefore, been focused on two case studies, in a Mixed Teak Forest System in Gujarat, and a Sal Coppice Forest System in West Bengal. The areas were carefully selected, in consultation with GOI and state Forest Departments, as representative of commonly occurring social and ecological conditions. The focused nature of the study means, however, that the results can only be extrapolated to other areas and forest systems with caution, though the study should still be of interest to planners and policy makers. A review of the literature showed that an appropriate analytical method was lacking, so another important objective of this study is to develop and test a suitable method. The study should, therefore, be of interest to researchers and those interested in applying the method to other areas. It is also thought to be of relevance to those concerned with the introduction of participatory processes and, in particular, community management of resources in other sectors.

## Analytical Method

The analytical method developed for this study is based on three closely linked models, developed from data collected in the field and from secondary sources. The main features of the models are summarized below and the linkages between them illustrated in Figure 1.

- The **Village Model** defines the socio-economic structure of the village and identifies the main stakeholders and their dependencies on associated forest areas. The main stakeholders are the nation or state, the Forest Department, and the community. Within the community there may be overlapping groups of stakeholders, such as wage earners, fuelwood collectors, known as headloaders, livestock owners, and collectors of Non-Timber Forest Products (NTFP). Some of these stakeholders presently exploit the forest, either legally or illegally, so that improved management and protection by the community restricts their access to forest products.

- The **Biological Model** estimates production from the forest, which includes timber at final harvest, and a range of products collected over a period of years, such as firewood, poles and other NTFPs, including grasses, fruits and leaves. The model estimates the biological growth of the forest, both with and without JFM, derived from measurement of present growing stock and projections of incremental growth and canopy cover in different forest types. Increase in the canopy cover with improved protection of the forest, affects the production of NTFPs, in some cases adversely. To estimate these changes, the model includes linkages between growing stock and canopy cover, and between canopy cover and NTFP production.

- The **Economic Model** simulates returns to JFM under different institutional and management conditions, represented by specific control variables. The economic model also identifies the returns to the different stakeholders. It is based on the valuation of inputs and outputs.

## Analytical Results

The economic model separates: (a) **economic returns** to JFM; (b) the **revenue sharing** between the Forest Department (FD) and the Forest Protection Committee (FPC), derived from timber and other outputs in the medium to long-term; (c) **income**, derived mainly from Non-Timber Forest Products (NTFPs) and wage employment, and; (d) its **distributional impact**. The results of the analyses show that the:

- **Economic returns** to JFM in both forest systems are considerable, as the net worth of the forest area under FPC management increases as a result of JFM. For the specific areas studied, net worth in the Mixed Teak Forest System increases by almost 60 percent, and by about 35 percent in the Sal Coppice Forest System. Most of the increase is due to better management of the natural forest. The contribution of NTFPs and other non-revenue flows, such as grazing, towards total net worth is significant;

# Figure 1.   The Analytical Model

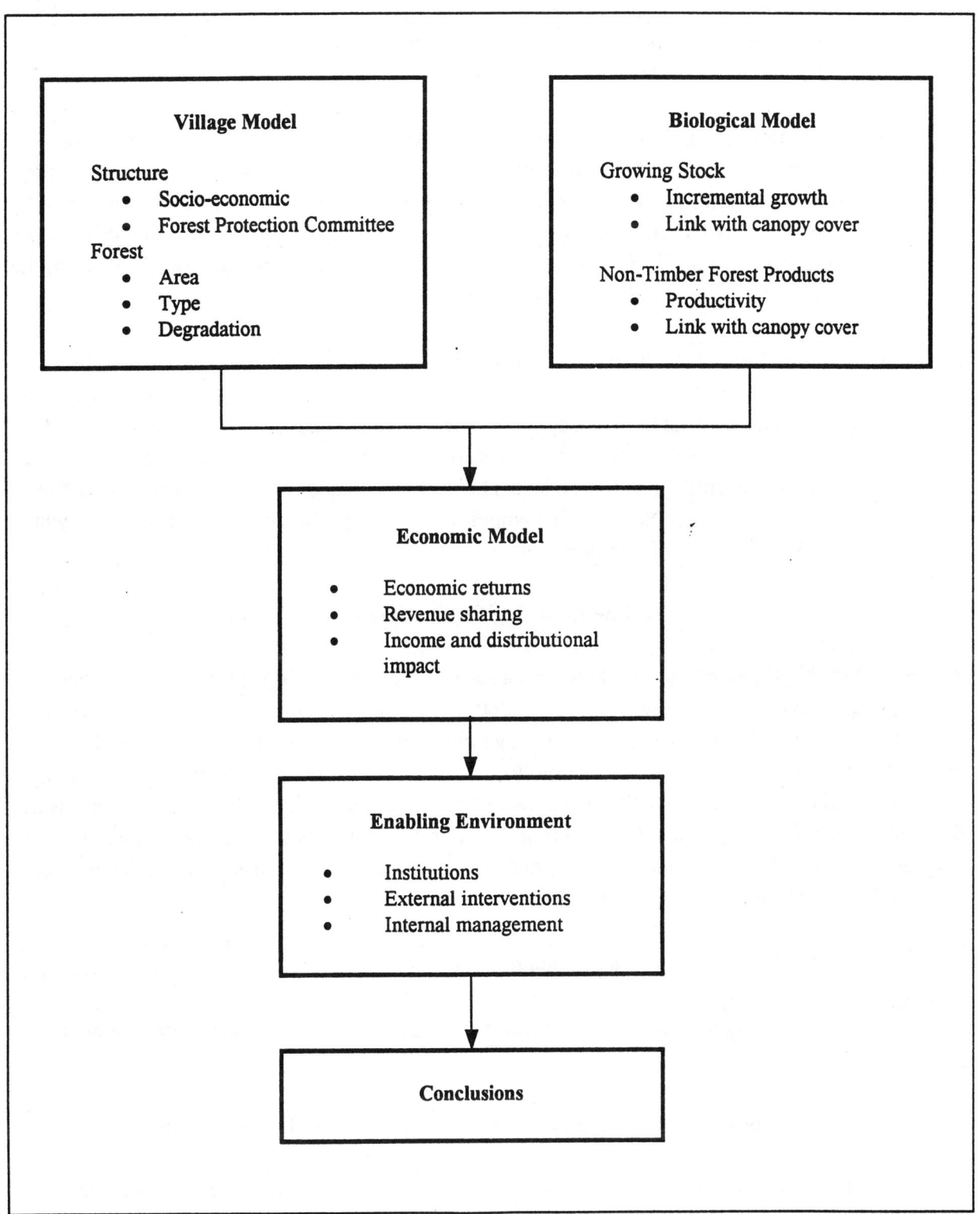

- **Revenue sharing**, determined by sharing arrangements specified in the State JFM Government Regulations, results in an incremental gain to the FPC in both forest systems as without JFM the FPC would not have received anything. The results also show apparent losses to the FD with JFM, but as the potential revenue without JFM is unlikely to be fully realized, due to uncontrolled, illegal, and premature exploitation, the FD is also likely to gain;

- **Income** flows into the FPC, with JFM, increase significantly in the Mixed Teak Forest System, reflecting the quality of the forest in the study area. Analysis of the returns to various stakeholders in the village shows that there are substantial overall gains for NTFP collectors. In the Sal Coppice Forest System the considerable gains in revenue are offset by a decrease of the FPC income, due mainly to a decrease in the production of tendu leaves as the canopy cover closes.

- The analysis of the **distributional impact** of JFM shows that in the Mixed Teak Forest System Model, where there is an overall increase in income, the main losers are headloaders. In the Sal Coppice Forest Model, where there is slight fall in total income, both headloaders and NTFP collectors are losers. However, there are sufficient overall gains for the losers to be compensated, either from revenue or from collection of other NTFPs. This emphasizes the importance of the full involvement of all groups in JFM decision-making

## The Enabling Environment

These findings highlight the fact that both social relationships and biological linkages between NTFP production and forest growth are site-specific and complex. In order to take account of this, a key element of participatory forest management must be to base decisions on local understanding and objectives, that can be modified in response to site specific experience. The ability of local communities to make decisions that take account of the values and perceptions of different stakeholders depends on an enabling environment that consists of complex institutional and social conditions. Consequently, the realization of the benefits discussed above depends on this enabling environment. In particular, it depends on:

- The representativeness and functioning of the FPC;

- The institutional and regulatory framework, in particular the sharing arrangements; and

- The regional economic context, for example, the availability of markets.

Because of the complex inter-relationships between these factors and site-specific ecological factors, it was not possible to take account of them in a quantified manner in the analysis. However, they are discussed in the light of data collected from a number of villages in the two study areas and information contained in the literature.

## Follow-Up

The results of the study depend to a large extent on the relationships and data included in the models. The data was collected in the field, but in a limited number of sites. Follow-up studies are, therefore, required and include:

- Application of the analytical method in a variety of ecological and social conditions to test the extent to which the conclusions can be extrapolated;

- Estimation of the annual increment in degraded natural forest, less than 10 years old, under different biotic pressures, in order to establish the relationships between Mean Annual Increment (MAI) and Current Annual Increment (CAI) and to calibrate the analytical model;

- Measurement of NTFP productivity in relation to forest degradation in a variety of ecological conditions;

- Assessment of fodder resources, both herbaceous and tree fodder, derived from forest of different levels of degradation;

- Investigation of the internal dynamics of FPCs, with particular reference to the representation of disadvantaged groups.

## Implications for the Future

Although the results of the study can only be extrapolated with caution, they suggest that the conclusions of earlier analyses undertaken during appraisal of Bank-supported forestry projects have been broadly correct and that Bank financing for JFM programs has been appropriate. However, the conclusions of the study with regard to the uneven distribution of benefits within the community, point to the need for future projects to include specific measures to safeguard the interests of those who may lose from the introduction of JFM, especially as they are often the poorest and most vulnerable in the community. The study is also relevant to Bank lending in other sectors in which community management of resources is being introduced, for example, community management of irrigation schemes. Thus, the analytical approach, and in particular, the need to consider the distributional impact of schemes that appear to benefit communities as a whole, have relevance to other sectors.

# I. INTRODUCTION

## Background

### Policy Framework

The National Forestry Policy of 1988 is to treat forests, first, as an ecological necessity, second, as a source of goods for use by local populations, with particular emphasis on non-timber forest products, and third, as a source of wood and other products for industries and other non-local users. The policy envisages participation of communities in the management of forest resources as a means of achieving these objectives, in particular, providing a sustainable system of management to avoid further the deforestation or degradation of forests on government and common lands. In June, 1990, the GOI Ministry of Environment and Forests (MOEF) issued policy instructions to all state forestry departments which defined an operational and institutional framework for the new policy. Thirteen state governments have already passed Government Orders whilst others are in the process of developing initiatives.

### Joint Forest Management

The participatory approach to management of forest resources is commonly referred to as Joint Forest Management (JFM). It can be defined as follows:

- *Joint Forest Management* is the sharing of products, responsibilities, control, and decision making authority over forest lands, between forest departments and local user groups, based on a formal agreement. The primary purpose of JFM is to give users a stake in forest benefits and a role in planning and management for the improvement of forest conditions and productivity. A second goal is to support an equitable distribution of forest products.

There is already an extensive literature on JFM in India and elsewhere. Much has been summarized in publications of the Society for Promotion of Wastelands Development, by the Tata Energy Research Institute, by Poffenberger and McGean, and in a recent World Bank Forest Sector Report [1]. The 1994 World Bank Sector Portfolio Performance Report indicates the contribution that JFM makes to the development of village institutions and participatory processes, and to the alleviation of poverty. There are, however, three main dimensions to the analysis and discussion of JFM; social and institutional issues, technical parameters, and economic factors. A review of existing literature undertaken for the present study showed that social and institutional arrangements have been described in detail in a wide variety of sites, but there has been little systematic discussion of technical forest management. Quantitative analysis of the financial and

---

[1]    Society for Promotion of Wastelands Development (SPWD). Bibliography on JFM. 1995. SPWD, New Delhi.
    Tata Energy Research Institute (TERI). 1992. Joint Forest Management Series. TERI. New Delhi.

    Poffenberger M. and McGean N. Eds. *Village Voices, Forest Choices. Indian Experience with Joint Forest Management.* Oxford University Press.

    World Bank. 1993. India. Policies and Issues in Forest Sector Development. Report No. 10965 - IN.

economic returns to JFM and, in particular, of the incentives for communities to participate is also lacking. From a review of over 100 studies of JFM only 41 presented information on financial and economic factors and few of these included any systematic analysis of costs and benefits. Analysis of the distribution of benefits within a community is also lacking, even though this has important equity and gender implications. Although models of the costs and benefits of JFM, that reflect the incentives for communities to participate in the management of forest resources have been developed, there has been no systematic attempt to develop an appropriate methodology, that takes into account the particular characteristics of JFM.

Successful implementation of JFM in Bank-assisted projects and elsewhere, coupled with the emphasis in the National Forest Policy on community participation in forest management has resulted in a considerable increase in the number of proposals for the introduction of JFM. Support for JFM forms an important part of existing Bank-funded forestry projects, and is likely to be an integral part of other future forestry projects. Given the likely expansion of JFM arrangements, there is an urgent need to validate JFM models in a variety of social and ecological conditions. This would provide a better understanding of the incentives for communities to participate in forest management and a firm analytical basis for preparation of future forestry projects.

## Study Objectives

The main objective of this study is to develop a better understanding of the incentives for communities to participate in JFM, in particular, the economic and financial incentives for participation of various groups of forest users, the stakeholders in the forest resources, both within and outside the community. In response to the National Forest Management Policy, 1988, JFM arrangements have been introduced in a wide range of ecological and social conditions in India. However, for the purposes of this study, it was decided to undertake detailed analysis in a limited range of conditions, rather than more general analysis over a wider range of conditions. The study has, therefore, been focused on two case studies, in a Mixed Teak Forest System in Gujarat, and a Sal Coppice Forest System in West Bengal. The focused nature of the study means, however, that the results can only be extrapolated to other areas and forest systems with caution, though the study should still be of interest to planners and policy makers. An appropriate analytical method was lacking, so another important objective is to develop and test a suitable method. The study should, therefore, also be of interest to researchers and those interested in applying the method to other areas. The final objective of the study is to examine the analytical results in the light of social, institutional and regulatory factors. Site specific field observations are related to a more general framework of experiences with JFM, as reflected in reviews of JFM and participatory arrangements presented in many other published papers. Consequently, the study is thought to be of relevance to those concerned with the introduction of participatory processes and, in particular, community management of resources in other sectors.

## Study Methodology

### Site Selection

The study is derived from work undertaken by a small team of consultants financed through the Netherlands Consultant Trust Fund in collaboration with World Bank staff. Study methods were discussed with a Working Group established by MOEF and it was agreed that a focused and detailed study in a limited number of sites was preferable to a more generalized study over a larger number of sites. The Working Group recommended sites be chosen in West Gujarat and West Bengal that would be representative of Mixed Teak Forest Systems and Sal Coppice Forest Systems, respectively. Specific locations in Gujarat, the Rajpipla (East) Forest Division, and in West Bengal, the Bankura (South) Forest Division, were selected following further discussion with the two State Forest Departments.

### Data Collection

Following a review of available literature, field studies were undertaken in both sites over a period of 8 weeks by a team consisting of a forester, an economist and a sociologist. After discussion with FD staff and others and using secondary data, representative Forest Protection Committees (FPCs) were selected for detailed study. Participatory appraisal techniques were used to build up an understanding of forest use and dependency, to identify sub-groups within the community and to select forest areas for a resource survey in order to estimate forest status and productivity. Apart from the main study villages, data was also collected in nearby villages. A wide range of sources at different points along the marketing chain, including villagers, official sources and traders, provided marketing information and price data.

### Presentation of Results

This report presents the analytical method developed for the study, the application of the method to villages in two forest systems, and a discussion of the results of the analyses. It is based on field studies and analyses undertaken by a team of consultants (See Annex 1).

## II. THE ANALYTICAL METHOD

The method is based on the development of three closely linked models, as illustrated in Figure 1: (i) the **Village Model** defines the socio-economic structure of selected representative villages, and identifies the main stakeholders and their dependencies on associated forest areas; (ii) the **Biological Model** estimates production of forest products, both with and without JFM, from calculations of biological growth, based on the growing stock and relationships between growing stock and canopy cover, and between canopy cover and the production of NTFP; (iii) the **Economic Model** simulates returns to JFM under different management situations, as represented by specific control variables. The economic model also identifies the returns to the different stakeholders. It is based on the valuation of inputs and outputs. In this chapter, the principles used to construct the models are discussed, followed by a consideration of the control variables. Analyses and models, based on data from representative village areas in two forest systems are presented in Chapter 3, and the results discussed in Chapter 4.

### The Village Model

Models were developed for selected representative village areas, based on field studies, as discussed in the following chapter. The socio-economic structure of the village and the present structure and status of the associated forest areas are site specific. However, in each village, three main groups of users of forest resources, the stakeholders, were identified, each of whom have different incentives to participate in JFM, as summarized in the following table:

| Stakeholders | Incentives |
|---|---|
| Nation, state, or society | Economic benefits |
| Forest Department | Departmental revenue |
| Community or Forest Protection Committee | Community revenue or produce |

Within a community there may be several overlapping groups of stakeholders;

- Revenue earners,
- Wage earners,
- Fuelwood headloaders,
- Livestock owners, and
- NTFP collectors

Each group uses forest areas for harvesting or collection, and these areas were identified and their status measured.

## The Biological Model

Projections of the biological growth of the forest system, with and without JFM are based on:

- The use of Current Annual Increment (CAI) [2], that is the marginal product, rather than Mean Annual Increment (MAI), the average product, to estimate growth in unmanaged forest (Annex 2). The use of CAI, or actual annual growth, rather than MAI, or average annual growth, provides a more exact way of simulating forest growth in the short term;

- Biological relationships which link productivity of NTFPs to forest growth. In some cases, such as for grazing and some NTFPs, these relationships were determined indirectly, by first linking growing stock to canopy cover and then, productivity to canopy cover;

- management and control variables, such as the offtake, or "hacking" rate [3], which ensures that changes in these variables resulting from the introduction of JFM are made explicit;

## The Economic Model

The change in net worth of the village forest as a result of the introduction of JFM is used to measure the impact of JFM. A discount rate of 12% [4], representing the opportunity cost of capital nationally, enables a comparison to be made with investments in other sectors. A planning horizon of 30 years is assumed.

JFM involves the official transfer of a stake in the forest from the state to the FPC. This stake includes the land and timber value of the forest, that is, its capital value, as well as income derived from the forest. It is distinct from a welfare program in which the community benefits from employment generation or subsidized asset creation. The analysis, therefore, separates:

- the *economic returns* to the stake in the forest. These are based on the "stumpage" values for timber and site values for NTFPs. These are net of collection or harvesting costs and represent economic values.

---

[2]     For simulation purposes, CAI was used as a percentage of current growing stock, referred to as the Current Increment Percent (CIP).

[3]     The term 'hacking' is taken to mean continuous harvesting and includes lopping of older trees as well as cutting of small trees, thereby reducing woody growth

[4]     The choice of the Opportunity Cost of Capital as the discount rate is only really applicable at the economic level where forestry investment is an alternative to other forms of state investment. For other stakeholders, there are likely to be different time preference rates (TPRs); poorer sub-groups are likely to have TPRs in excess of the national OCC. However, data on TPRs for rural subgroups was not available. Conversely, the models could be developed to reveal TPRs for sub-groups.

- the *revenue effect* of sharing of outputs between the Forest Department and the FPC;

- the *income effect* of JFM including both the stumpage and wage value of products. The returns are based on the value of the forest products at the point of use or sale and, therefore, include the value of labor involved in collection and harvesting. This will determine individual perceptions of the potential impact of JFM.

## Management or Control Variables

The impact of JFM depends on three groups of management or control variables: (a) the institutional framework; (b) external interventions and; (c) internal management for protection and management of forest resources. Changes in these variables are reflected for both the "with" and "without" JFM situations. In order to estimate the benefit flows associated with these variables, empirical proxies were identified as follows:

| Variable | Proxy |
|---|---|
| (a) Institutional framework | Revenue sharing between FD and FPC |
| (b) External interventions | Planting |
| (c) Internal management | Protection - level of hacking |

## Institutional Framework

The current arrangements for revenue-sharing between the FD and FPC as stated in the Government Orders and Resolutions, are as follows:-

| State | Final Harvest | Intermediate Thinnings | NTFP |
|---|---|---|---|
| West Bengal | 25% of final produce | 100% of intermediate products | Collection:<br>  - villagers<br>Marketing:<br>  - LAMPS[5] /FDC<br>  - Traders |
| Gujarat | 50% of final produce to FPC | 0% of intermediate products | Collection:<br>  - villagers<br>Marketing:<br>  - LAMPS/FDC<br>  - Traders |

In the situation without JFM, that is, without the formation of an FPC, the community would have no legal share in the final harvest or thinnings.

---

[5]     LAMPS. Large-Sized Multi-Purpose Cooperative Societies.

## External Interventions

JFM has largely been introduced in association with Forest Department planting programs, both plantations and enrichment planting in natural forest, which mean that there are immediate employment benefits. This helps to compensate for the loss of income suffered by some members of the community, due to the improved protection associated with JFM, that reduces illegal exploitation of forest land. In the establishment phase, planting provides employment in areas with few alternative employment opportunities. Apart from enhancing the overall value of the forest area, planting changes the mix of species and, thereby, the products, yields and harvesting times. The resulting changes in the flow of benefits to the village are taken account of [6] in the analysis.

## Internal Management

The FPC is responsible for the protection of forest areas under their control. The management, or level of protection is measured by the rate of "hacking" as a percent of CAI, estimated on the basis of differences between similar forests in nearby areas. With JFM, forest management improves and the "hacking" rate is assumed to fall.

Initially, the FPC will focus on protecting the village area from exploitation by outside groups, followed by the development of mechanisms for internal regulation, the rules and norms for management and use of the area by the community itself. The success of internal regulation depends on the supply of forest produce, both inside and outside the village, and on the capacity of the village to develop and maintain internal management rules. In practice, new plantings are generally protected more effectively than natural forest. However, irrespective of the regulatory framework, it is expected that the community will meet their domestic consumption from local supplies or, where this is not possible, from non-assigned nearby forests.

## Environmental Values

Environmental benefits associated with better quality forest have not been included in the model. However, with effective protection under JFM, the likelihood of the realization of these benefits is increased.

---

[6]     As well as planting, the FD has also implemented support activities, associated with JFM but with funds from other schemes, in order to increase the incentives to communities to participate in JFM. These activities provide immediate employment as well as creating assets. The impact of these activities is difficult to quantify as data is not readily available.

# III. APPLICATION OF THE ANALYTICAL METHOD

## General

The principles described in the previous chapter were used to develop a model for each of two forest systems, based on villages selected in discussion with MOEF and the State Forest Departments in Gujarat and West Bengal. In each case, the village model is presented, followed by a discussion of the biological model. The economic model, based on valuation of inputs and outputs is then discussed and the data used to simulate the returns to JFM. Data used in the models were collected in the field or from the best available secondary sources. More complete descriptions of the study areas are provided in Annexes 3 and 4. The results of the analyses are presented and discussed in Chapter 4.

## The Mixed Teak Forest System - Gujarat

### The Village Model

**Population and Socio-Economic Structure.** The model for this system was based mainly on data from the village of Dhanor in Rajpipla Division, Gujarat. There are 56 households (290 people) living in the village; they belong to the Vasava clan of the Bhil tribe. Five of these households are landless, but the remaining households have access to undivided portions of land registered in the names of older family members. A mixed crop including paddy, jowar, tuvar and coarse grains is grown under rainfed conditions. Agricultural labor is the main source of income for most families; only five households are self-sufficient in food production. After the agricultural season, many migrate to find work on construction sites. Some of the youths work in Surat and only return for a few days each month.

**Forest Resources.** The village has approximately 300 ha of forest land including 8 ha of grazing land (Table MT 1) [7]. Just over 205 ha has been handed over to the FPC, the remainder is still under FD management. It has been assumed that this land will not be affected by JFM, as it is situated some distance from the village and there is no data to predict likely future use of the area. Although there is a large sanctuary close to the village, there is also considerable degradation occurring throughout the area and, even in well-stocked areas, the average age of the forest is low, due to past heavy felling by FD, and local dependence on the forest. The field survey showed that the forest land under JFM in the selected village can be considered under four categories of natural forest (NF 1-4, Tables MT2 and MT3) and five categories of plantation, depending on species and age of planting (PL 1-5, Table MT2). There is also village land set aside for grazing, whilst the remainder of the natural forest is managed entirely by the FD.

All households collect fuelwood for domestic consumption, timber for agricultural implements and house repairs. All households cut and carry grass either for their own use or for sale or both. All households use the forest for grazing and collecting NTFPs. Roughly 60% of households also headload fuelwood for sale. The main wood products were timber, poles and fuelwood. The main

---

[7]    Tables relating to the Mixed Teak Forest System are given an MT prefix.

species was teak; the other species were grouped together using a weighted average based on volume. Important NTFPs include tendu and fruit.

An FPC was established as a Van Samiti in 1990/91 by the FD. Forty three households are members of the FPC, which is functioning well; there is strong leadership and good support. However, there are also conflicts, in particular, from a group of five or six unemployed people who headload for sale.

## The Biological Model

**Management Practices.** The assumptions used were as follows:

| | | |
|---|---|---|
| With JFM: | (a) | degraded areas without sufficient rootstock will be replanted; |
| | (b) | degrading areas with rootstock will be protected; |
| | (c) | overstocked areas will be managed to achieve optimal stocking. |

| | | |
|---|---|---|
| Without JFM: | (a) | degraded areas without rootstock that will remain degraded; |
| | (b) | in degrading areas teak will be hacked on an annual basis, without allowing any shoots to prosper; |
| | (c) | overstocked areas will be degraded. |

**Forest Productivity.** The main forest products include wood, in the form of timber, poles and fuelwood, bamboo, grazing, NTFPs and fruits. Productivity models for both natural forests and plantation areas are presented in Tables MT 4-7 and MT 8-12, based on field measurements of sample plots. Assumptions about biological relationships, used in the models are discussed below.

**Wood - Natural Forests.** Productivity of wood in natural forests was estimated as a function of current growing stock, CAI, and offtake:

- *Growing Stock* was calculated from measurements of stems, basal area, height, girth and age in sample plots in both natural forests and planted (Table MT3);

- *CAI* in each year was derived from standard yield tables and used to calculate final yields (Table MT 13);

- *Offtake* rates were estimated on the basis of comparisons between different forest areas and estimates of demand derived from village surveys (Table MT 14). Offtake rates in excess of 100% represent destruction of growing stock. As well as any green cutting, dead and dry material was assumed to be available each year at 10% of the CAI for the previous year.

Projected yields from natural forests are summarized below and presented in more detail in Table MT 15.  Final GS represents biomass in the form of timber and poles, and excludes intermediate products, collected through hacking.  Final GS with JFM is higher than without JFM.  However, total yield, which includes the intermediate products, is higher without JFM, though it is predominantly composed of small stock.

| Natural Forest Type (Table MT3) | Protection | Offtake % | Year 50 | |
|---|---|---|---|---|
| | | | Projected Final GS m$^3$/ha | Total Yield m$^3$/ha |
| NF 1 | With JFM | 10 | 58.8 | 58.8 |
| | Without JFM | 30 | 56.4 | 63.1 |
| NF 2 | With JFM | 10 | 57.3 | 57.3 |
| | Without JFM | 50 | 49.0 | 70.3 |
| NF3 | With JFM | 10 | 60.8 | 60.8 |
| | Without JFM | 90 | 38.1 | 106.6 |
| NF4 | With JFM | 10 | 60.3 | 60.3 |
| | Without JFM | 120 | 0.0 | 129.7 |

**Wood - Plantations**.  Productivity of wood from plantations was estimated using productivity in existing natural forest as shown below.

| Species | MAI m$^3$ per 1000 stem per ha |
|---|---|
| Teak | 15 |
| Sawan/Sadar | 10 |
| Non-teak mix | 3 |
| Kakad | 5 |
| Stylozanthes | 5  (MT/ha) |

Stems, basal area, height, girth and age for sample areas of plantation forests were measured to calculate GS. Protection is generally effective in plantation areas so offtake was assumed to be minimal. It was also assumed that a mix of species would be planted as summarized below.

| Plantation Type (Table MT 2) | Year | Area(ha) | Species |
|---|---|---|---|
| PL1 | 1 | 2.5 | Teak |
| PL2 | 2 | 5 | Sevan/Sadar |
| PL3 | 3 | 40 | Subabul/SWC/MFP |
| PL4 | 4 | 10 | Fodder |
| PL5 | 5 | 50 | Waste/Mixed |

Note: SWC - Soil and water conservation     MFP - Minor Forest Products

**Bamboo**. With JFM, improved management and regular harvesting increase the quantity and quality of bamboo. The assumed increases are as follows:

| Bamboo | Age | | | |
|---|---|---|---|---|
| | 1 | 4 | 8 | 12 and every 4th year |
| Harvest (culms/clump) | 2 | 3 | 6 | 8 |

Source: Standard Yield Tables

Without JFM, the same growth rate is assumed, but with indiscriminate hacking, the annual yield is 2 culms per clump, most of which is of lops and tops quality. This quality difference is reflected in the value of the crop.

**Canopy Cover (CC)**. An increase in the canopy cover, linked to an increase in tree biomass, as a result of JFM, has implications for the availability of grazing and other forest products. The relationship between CC and tree biomass depends on species, basal area and age and is not uniform over time and is not normally recorded. Despite this difficulty, and on the basis of ocular estimates of CC, the following relationship between canopy cover and growing stock was estimated:-

$$CC = 0.188 + 0.424 \times (GS) + 0.0152 \times (GS)^2 \qquad R^2 = 0.99$$

The formula provides a means of estimating changes in CC as a result of growth or degradation in the growing stock (Table MT 16).

**Grazing**. The availability of grass for grazing by animals is a function of CC, and other factors, in natural forest. The quantity of grass available by regular grazing is difficult to measure. The standard assumption is that without any CC (CC =0), but with forest rootstock, the grazing potential is equivalent to 1.6 tons. By the time the CC has risen to 50%, the potential has reduced to browsing and tree fodder, estimated to be 0.60 tons (Table MT 17). A simple linear relationship between canopy cover and grazing yields is assumed.

$$\text{Grazing} = 1.6 - (0.02) \times CC\%, \text{ over the range } CC\%: 0 \text{ to } 50$$

Beyond this range, the grazing potential is assumed to remain constant at 0.60 MT. Grass yields can be much higher that these estimates; where the land has no forest root-stock, grazing productivity may rise to over 2 MT per ha. Without protection, the soils are compacted and grass yields decline; the impact of compaction is not quantified, but the value of grass from protected and unprotected plots has been estimated and used to differentiate between the two outputs. In plantations, initial grass yields rise as a result of soil working and then decline over time, as the canopy closes over.

There is little empirical data available so estimates were based on the average of estimates from two local organizations. The assumptions used in the study are presented below:

| Grass Yield (tons/ha) in Plantations by Year | | | | | | | | | |
|------|------|------|------|-----|-----|-----|-----|-----|-----|
| 1 | 2 | 3 | 4 | 5 | 6 | 7 | 8 | 9 | 10 |
| 1450 | 1200 | 1200 | 1100 | 750 | 650 | 650 | 650 | 500 | 400 |

**NTFPs**. Tendu leaves are collected from both bushes in degraded areas and from trees in less degraded areas; bushes are more prolific where the canopy is more open. A linear relationship between the value of leaves collection per ha and CC was assumed. The value of other NTFP collected was based on data from informants. At low CC, there was no NTFP collection, However, in very dense forest, the NTFP collection also fell; the maximum appeared to be about 60% canopy cover. There was little consensus; ideal conditions differed from product to product and also depended on factors other than CC. On the basis of these discussions, the following relationships were assumed.

| Canopy cover (%) | Tendu Rs/ha | NTFP (Rs/ha) | |
|---|---|---|---|
| | | Natural Forest | Plantations |
| 0 | 120 | 0 | 0 |
| 10 | 100 | 25 | 0 |
| 20 | 80 | 50 | 0 |
| 30 | 60 | 150 | 75 |
| 40 | 40 | 250 | 150 |
| 50 | 40 | 350 | 225 |
| 60 | 40 | 450 | 300 |
| 70 | 40 | 450 | 300 |
| 80 | 40 | 400 | 300 |
| 90 | 40 | 350 | 300 |
| 100 | 40 | 300 | 300 |

**Fruit**. Amla was taken as a proxy for a wide range of different fruits. Yield assumptions were as follows:

| | Age | | | | | |
|---|---|---|---|---|---|---|
| | 1-6 | 7-10 | 11-15 | 16-20 | 21-25 | 26-30 |
| Yield (kg/tree) | 0 | 1 | 10 | 20 | 30 | 40 |

## The Economic Model

**Wage Rates**. Within the village wage rates for unskilled daily agricultural labor are Rs 12 per day plus a meal and bidis, estimated to cost about Rs 3, giving a mean rate of Rs 15 per day. This was used to value labor, except for employment in plantations in which FD rates of Rs 39 per day were used.

**Plantation costs**. Unit plantation costs are based on schemes used in the study area and are typical of costs used elsewhere in India. Virtually all costs are for labor. As the material costs, including the costs of seedlings and small quantities of fertilizer amount to only about 5 percent of total costs, they have not been specified in detail.

| Item | Planting Costs (Rs/ha) by Year | | | | | | Total |
|---|---|---|---|---|---|---|---|
| | 0 | 1 | 2 | 3 | 4 | 5 | |
| Labor | 420 | 6,670 | 2,390 | 780 | 160 | 160 | 10,580 |
| Materials | 290 | 160 | 240 | | | | 690 |
| Total | 710 | 6,830 | 2,630 | 780 | 160 | 160 | 11,270 |

Source: FD Gujarat.

**Prices**. The prices of forest products at different points in the marketing chain were collected and stumpage or field values determined by adjusting for the processing and labor costs. A summary of the prices (financial and economic) used to value products is given below. More details are presented in Annex 5.

| Product | Unit | Financial Market Prices | Economic 'Stumpage' Prices |
|---|---|---|---|
| Teak    Wood | $m^3$ | 11,500.00 | 8,324.00 |
|        Poles -small | pole | 4,020.00 | 2,699.00 |
|        Poles -large | pole | 4,230.00 | 2,844.00 |
|        Fuelwood | $m^3$ | 1,680.00 | 1,156.00 |
| Non-teak wood | $m^3$ | 6,100.00 | 4,272.00 |
| Fuelwood headloaded | qtl | 135.00 | 108.00 |
| Bamboo Good qlty | culm | 7.55 | 8.00 |
|        Mixed qlty | culm | 1.55 | 2.00 |
|        Lops/tops | kg | 0.11 | 0.11 |
| Grazing | ton | 460.00 | 410.00 |
| Tendu | Rs/ha | 33.60 | 33.60 |
| Fruit | kg | 2.50 | 2.50 |

**Price Trends**. The value of forestry products depends on future, rather than current, prices. Current prices have been taken as proxies for future prices; these probably best reflect FPC and FD perceptions of value and are therefore appropriate in an analysis of incentives.

**Benefit Flows**

For each type of natural forest (Para. 3.3) the flow of benefits with and without JFM was simulated on a per hectare basis and the data presented in Tables MT 4-7. The difference between the with and without JFM situations is due to different assumed offtake rates (See Para. 3.7). Based on the areas of different types of natural forest in the village, the incremental flow of benefits from natural forests for the village as a whole was calculated, and is presented in Table MT 18.

Similarly, the flow of benefits from plantations was estimated, both with and without JFM. In Tables MT 8-12 the benefits with JFM are presented for each plantation type (PL 1-5, Para 3.3) based on the actual areas of plantations in the village area, productivity estimates (Para. 3.9) and plantation costs (Para. 3.18). The without JFM situation is assumed to follow the same pattern as degraded natural forest (Tables 4-7). These data are also included in Tables MT 8-12 for ease of reference, which present the incremental flow of benefits from plantations with JFM. Total benefit flows from both natural forest and plantations are presented in Table MT 19. The conclusions to be drawn from the model are discussed in Chapter 4.

## The Sal Coppice Forest System - West Bengal

### The Village Model

**Population and Socio-Economic Structure.** The model for this system was based mainly on data from the village of Fakirdanga in Bankura South Division, West Bengal. There are 111 households living in the village, mostly tribal people of whom the largest group are Santhals. Sub-groups within the village, usually based on tribe and caste, live together in their own hamlets. Cultivation of mixed crops is mainly rainfed and livestock are kept mainly for draft power and manure. Agricultural labor is the main source of income for most families.

**Forest Resources.** The forest area totals about 130 ha, of which 50 ha are plantations, 7 ha are encroached and the remainder natural forest, that can be divided into highly degraded (23 ha), moderately degraded (20 ha) and undegraded (30 ha) Table SC 1 [8]. About 7 ha of forest land have been encroached on and are used for agriculture. Field surveys showed that the forest land under JFM can be considered in three categories of natural forest (NF1-3, Tables SC 2,3), and two categories of plantation, depending on species (PL1-2, Tables SC 2,3).

Almost all households collect leaf litter and fuelwood from the forest. The major impact of leaf litter collection is to reduce the recycling of biomass, resulting in reduced tree and NTFP productivity. All households obtain timber for agricultural implements and house repairs from the forest. About 30 percent of households are thought to be involved in charcoal making, though data on quantities produced were not available. All households in the village are members of the Forest Protection Committee. People's participation in forest protection was introduced by the FD in response to the failure of traditional methods of forest control. Awareness of the need for popular support for forest protection has resulted from the rapid period of degradation since nationalization of Zamindari forests in 1954. This awareness has been focused because of the commitment to people's involvement by all the political parties. Within the FD, there has been active support and leadership for people's participation and JFM by the Staff Association of the FD staff who saw JFM

---

[8]     Tables relating to the Sal Coppice Forest System are given an SC prefix.

as being in the interests of both members and society as whole.  There are few activist NGOs within the area and none in the study village.

**The Biological Model**

**Management Practices**  The assumptions used were as follows:-

With JFM it is expected that:

- degraded areas without sufficient rootstock will be replanted;
- degrading areas with rootstock will be protected;
- overstocked areas will be managed.

Without JFM, it is assumed that:

- degraded areas without rootstock will remain degraded;
- in areas that are currently being degraded, sal coppice is hacked on an annual basis without allowing any shoots to prosper;
- in the overstocked areas the rotation is extended beyond the prescribed 10 years.

**Forest Productivity**.  The main forest products include wood, primarily sal timber and poles. NTFPs are a major economic resource for both consumption and income and include sal leaf plates, sal seeds, mushrooms, tendu leaves and tussar cocoons.  Productivity models for both natural forests and plantations are presented in Tables SC 4,5,6, based on field measurements of sample plots.  Assumptions about biological relationships used in the models are discussed below.

**Wood - Natural Forests**.  Productivity of wood in natural forests was estimated in a similar way to that for the Mixed Teak forest system, as a function of current growing stock, CAI, and offtake:

- *Growing Stock* was calculated from measurements of stems, basal area, height, girth and age in sample plots in natural forests and planted areas (Table SC 3);

- *CAI* was derived from standard yield tables and used to calculate final yields (Table SC 8);

- *Offtake* rates were estimated on the basis of comparisons between different forest areas and estimates of demand derived from village surveys Table SC 9.  As well as any green cutting, dead and dry material was assumed to be available each year at 10% of the CAI for the previous year.

Wood productivity in this coppice forest system is estimated as follows. With JFM and effective protection and management, the current CAI of natural forest is accumulated and taken as multiple-shoot cuttings (MSC) and final harvest (Table SC 10). A ten year rotation is assumed, with yields as shown below:

| Year | Harvest | Yield (m³/ha) |
|---|---|---|
| 0 | Initial coppice | |
| 4 | MSC -I | 7 |
| 7 | MSC -II | 10 |
| 10 | Final coppice | 51 |
| | Total | 68 |

Productivity, without JFM, is more difficult to estimate. The traditional analytical approach is to used the MAI over a full rotation. However, this measure is unsuitable for annual cutting since the MAI will understate the yield. In this study and despite the limitations of data, an attempt was made to use CAI, expressed as percent of growing stock, a ratio known as Current Increment Percent (Annex 2).

**Wood - Plantations.** There are two areas of plantation, one Eucalyptus, the other Acacia, associated with JFM. Projected yields are based on measured yields at 10 years of age. Without JFM, the plantation areas would remain as highly degraded land with productivity estimated to be the same as for the most degraded areas of natural forest.

**NTFPs - Mushrooms.** Mushroom production is related to leaf litter availability. The quantity of mushrooms collected at present was estimated and apportioned to different forest areas on the basis of leaf litter collection production (Table SC 11). This showed a range in mushroom collection from 220 kg/ha/yr in well protected forest with 100 percent canopy cover to less than 80 kg/ha/yr in degraded forest areas. The relationship between mushroom production and growing stock is represented by the formula:

$$Mushrooms = 51.48 + 2.11 \times GS$$

This had to be developed indirectly, through the relationships between mushrooms and leaf litter availability, leaf litter and basal area, and basal area and growing stock, as shown in Table SC 11. The same relationships were applied to the without JFM situation.

**Tendu Leaves.** The production of tendu leaves is inversely related to canopy cover. However, there was insufficient data and returns per hectare were too low to model this effect. Total current production was distributed in the proportion 2:1 between highly degraded and moderately degraded forest areas; fully stocked forests were assumed not to provide any leaves.

Mushroom and tendu production account for only about 40 % of total estimated NTFP production. The relationship between these other NTFPs and growing stock was not estimated, since it became clear from PRA discussions, that the biological relationships were very complex and site-specific. Consequently, the same values for these other NTFPs had to be assigned to both the with-JFM and without-JFM cases. The model, therefore, includes an implicit assumption, obviously invalid, that JFM does not affect the bulk of NTFP production. The estimated values used were as follows:

| NTFP | Rs/ha |
|---|---|
| Sal plates | 514 |
| Sal seeds | 270 |
| Tussar | 393 |
| Miscellaneous | 885 |
| Sub-total | 2,062 |
| Mushroom | 936 |
| Tendu | 340 |
| Total | 3,338 |

**The Economic Model**

**Wage Rates.** Within the village, wage rates for unskilled daily agricultural labor range from Rs 10 to Rs 15 per day including a meal and bidi. Outside the village, general agricultural rates rise at peak times to about Rs 18, including extras, while labor working in nearby factories and coal mines earns considerably higher sums. FD rates for daily labor are fixed at Rs 39 per day and were used for plantation costs.

**Plantation Costs.** Planting costs for both natural forest and plantations were based on actual costs incurred by an existing project in the district. These costs are summarized below.

| Item | Planting Costs (Rs./ha) by Year | | | | | | | | Total |
|---|---|---|---|---|---|---|---|---|---|
| | 0 | 1 | 2 | 3 | 4 | 5 | 6 | 7 | |
| Sal Forest Regeneration | | | | | | | | | |
| Labor | 693 | 120 | 120 | 654 | 120 | 120 | 120 | 812 | 2759 |
| Materials | 10 | | | 10 | | | | 10 | 30 |
| Total | 703 | 120 | 120 | 664 | 120 | 120 | 120 | 822 | 2789 |
| Eucalyptus/Acacia | | | | | | | | | |
| Labor | 1600 | 4320 | 120 | 120 | 120 | 120 | | | 6400 |
| Materials | 160 | 30 | | | | | | | 190 |
| Total | 1760 | 4350 | 120 | 120 | 120 | 120 | | | 6590 |

**Prices**.  As with the Mixed Teak forest system, the prices of forest products at different points in the marketing chain were collected and stumpage or field values determined by adjusting for the processing and labor costs.  A summary of the prices used to value products is given below.  More details are presented in Annex 5.

| | Product | Unit | Financial Prices | Economic/ Stumpage Prices |
|---|---|---|---|---|
| Sal | Fuelwood | Rs/m$^3$ | 385 | 277 |
| | Poles (7) | Rs/m$^3$ | 1,388 | 1,211 |
| | Poles (7/10) | Rs/m$^3$ | 1,638 | 1,477 |
| | Poles (10) | Rs/m$^3$ | 1,763 | 1,610 |
| | Poles (16) | Rs/m$^3$ | 2,305 | 2,204 |
| | Poles (20) | Rs/m$^3$ | 1,957 | 1,891 |
| Eucalyptus | Fuelwood | Rs/m$^3$ | 375 | 239 |
| | Biomass | Rs/m$^3$ | 467 | 369 |
| NTFP | Mushrooms | Rs/kg | 12 | 9 |
| | Tendu | Rs/std bag | 200 | 130 |
| | 'Other' | Rs/ha | 2,061 | 699 |
| Wage rates | FD | Rs/day | 40 | 28 |
| | Agricultural | Rs/day | - | 15 |

**Price Trends**.  The value of forestry products depends on future, rather than current, prices. Current prices have been taken as proxies for future prices; these probably best reflect FPC and FD perceptions of value and are therefore appropriate in an analysis of incentives.

**Benefit Flows**

For each type of natural forest (Para. 3.25) the flow of benefits with and without JFM was simulated on a per hectare basis and the data presented in Tables SC 4,5,6.  The difference between the with and without JFM situations is due to different assumed offtake rates (See Para. 3.29). Based on the areas of different types of natural forest in the village, the incremental flow of benefits from natural forests for the village as a whole was calculated, and is presented in Table SC 12.

Similarly, the flow of benefits from plantations was estimated, both with and without JFM. In Table SC 7 the benefits with JFM are presented for each plantation type, based on the actual areas of plantations in the village area, productivity estimates, and plantation costs. The without JFM situation is assumed to follow the same pattern as degraded natural forest (Tables SC 4-6). Total benefit flows from both natural forest and plantations are presented in Table SC 13.  The conclusions to be drawn from the model are discussed in Chapter 4.

# IV. ANALYTICAL RESULTS

## The Mixed Teak Forest System - Gujarat

### Economic Returns

The results of the economic analysis are summarized below, from details presented in Table MT 19.

| Mixed Teak Forest System | Net Worth (Rs '000, 12%,30 yrs) | | | |
|---|---|---|---|---|
| | With JFM | Without JFM | Increment | (%) |
| Natural Forest | 3,875 | 986 | 2,889 | 293 |
| Plantation | 1,997 | 2,729 | - 732 | - 27 |
| Overall | 5,872 | 3,715 | 2,157 | 58 |
| % from Natural Forest | 66 | 26 | | |
| per ha (300ha) | 20 | 12 | 7 | |

These data show that the net worth of the forest area under FPC management increases as a result of JFM. Most of the increase is due to better management of the natural forest. The decline in net worth of plantations with JFM is due to differences in the way the plantations are managed. Without JFM, plantations are managed for timber with a higher economic value than the range of forest produce that would be realized with JFM. The current net worth of the total FPC managed forest area, in economic terms, using a discount rate (DR) of 12% is estimated to be Rs 3.71 million or Rs 12,000 per ha. This reflects the age and current level of stocking. With JFM and protection, the net worth might rise by 58% to Rs 5.9 million, or Rs 20,000 per ha. The natural forest area contributes 66% of the total with effective protection under JFM but less than 30% without JFM.

The following table shows that the contribution of NTFP and other non-revenue flows (grazing) towards total net worth is significant and, overall, rose by 105% as a result of JFM. Also, as a proportion of total net worth, these flows increase in importance from 26% of total net worth to about 34%. However, in the natural forest area, although there is a small rise (8%) in the value of NTFPs as result of protection, the contribution of NTFP to total net worth falls from 39% of net worth to only 11%. The plantations are largely for NTFP purposes and, therefore, contribute most of the non-revenue benefits.

| Mixed Teak Forest System | Net Worth (Rs.000, 12%, 30 yrs) | | | |
|---|---|---|---|---|
| | With JFM | Without JFM | Increment | % |
| Natural forest<br>- non-revenue flows<br>% non-revenue flows | 3,875<br>413<br>11 | 986<br>383<br>39 | 2,889<br>30 | 293<br>8 |
| Plantation<br>- non-revenue flows<br>% non-revenue flows | 1,997<br>1,575<br>79 | 2,729<br>589<br>22 | - 732<br>986 | -27<br>167 |
| Overall<br>- non-revenue flows<br>% non-revenue flows | 5,872<br>1,988<br>34 | 3,715<br>972<br>26 | 2157<br>1016 | 58<br>105 |

Note: Non-revenue flows include grazing as well as Tendu and other NTFPs.

## Revenue Sharing

Revenue from timber and bamboo harvest is shared between the FPC and FD. Intermediate products are not shared by the FPC. All other outputs, contributing to overall income flows, are assumed to belong to the community. The results of the analysis, presented below, show that with JFM, the FPC share of harvest revenue over 30 years amounts to Rs. 2.2 million, which is all incremental. The results also show apparent losses to the FD with JFM, but the potential revenue without JFM is unlikely to be fully realized for a variety of reasons, as discussed in Chapter 5.

| Mixed Teak Forest System | Net Worth (Rs.000, 12%, 30 yrs) | | | |
|---|---|---|---|---|
| | With JFM | Without JFM | Increment | % |
| Economy | 5,872 | 3,715 | 2,157 | 58 |
| FD Revenue | 1,641 | 2,283 | -642 | - 28 |
| FPC Revenue | 2,242 | 0 | 2,242 | - |

## Income and Distributional Impact

Overall there is an threefold increase in the value of income flows into the FPC from Rs 33,000 per household (equivalent to Rs 2,730 per annum into perpetuity at 12%) to Rs 110,000 per household (equivalent to annual income into perpetuity of Rs 9,300 at 12% or over Rs 600 days of agricultural work per year). This reflects the quality of the forest in the study area.

Stakeholders within the community have been identified in Chapter 3 and the value of returns to the various groups is presented below [9].

| Mixed Teak Forest System | Net Worth (Rs.000, 12%, 30 Years) | | | |
| --- | --- | --- | --- | --- |
| | With JFM | Without JFM | Increment | % |
| FPC revenue | 2,242 | 0 | 2,242 | - |
| Employment | 562 | 0 | 562 | - |
| Headloaders | 108 | 583 | - 475 | - 81 |
| Livestock owners | 611 | 608 | 3 | 0 |
| NTFP collectors | 1,270 | 240 | 1,029 | 429 |
| Net in kind | 2,551 | 1,431 | 1,120 | 78 |
| Total FPC income | 4,793 | 1,431 | 3,362 | 235 |
| Revenue per HH (43HH) | 111,500 | 33,250 | 78,250 | |

The figures show that there are substantial overall gains for NTFP collectors the value of whose income is projected to rise by over 400 %. Most of this is due to the NTFPs in the plantation areas. However, the headloaders are the main losers from increased protection, suffering a fall in income of about 50 %. This group are often the poorest in the village. The losses suffered by this group could be compensated for by employment on planting in the short run and by the FPC's share of revenue in the longer term, as discussed further in Chapter 5. The impact of protection on grazing is neutral in this case.

## Management or Control Variables

The impact of different assumptions about the revenue sharing, areas planted, and FPC management practices related to offtake, or hacking rates, was analyzed.

---

[9] The analysis of the distribution of benefits within the community depends on the income effect of the transfer of assets and was, therefore, carried out using market prices based at the point of sale/purchase of inputs and outputs.

**Revenue Sharing**. In Gujarat, revenue is shared 50:50 between FPC and FD. The impact of other sharing arrangements is summarized below. The switching value below which the FD incremental revenue becomes negative was estimated to be 65%.

|  | Net Worth (Rs.000, 12%, 30 Years) | | |
|---|---|---|---|
|  | With JFM | Without JFM | Increment |
| FD   (50%) | 1,641 | 2,283 | - 642 |
| FPC  (50%) | 2,242 | 0 | 2,242 |
| FD   (25%) | 519 | 2,283 | - 1,763 |
| FPC  (75%) | 3,363 | 0 | 3,363 |
| FD   (75%) | 2,762 | 2,283 | 479 |
| FPC  (25%) | 1,121 | 0 | 1,121 |

**Planting**. The overall rate of return to plantations, based on the planting schedule and allowing for planting costs but excluding FD overheads is 7%. An increase in the area under plantation will, therefore, reduce the overall net worth, discounted at 12%, of the village forest. However, there will be benefits in terms of employment generation and income during the period before harvesting.

**Offtake Rates**. The impact of hacking rates on net worth was analyzed and found to make little difference except where the quantity of timber or poles to be harvested at the end was reduced. This re-emphasizes the trade-off between current offtake, usually in kind, and revenue from the final harvest.

| Hacking Rate (%) Natural Forest With JFM | Net Worth (Rs.000, 12%, 30 Years) | | | |
|---|---|---|---|---|
|  | With JFM | Without JFM | Increment | % |
| 0 | 5,872 | 3,715 | 2,148 | 58 |
| 100 | 6,068 | 3,715 | 2,353 | 63 |
| 200 | 6,195 | 3,715 | 2,480 | 67 |
| 300 | 6,251 | 3,715 | 2,536 | 68 |

## The Sal Coppice Forest System - West Bengal

### Economic Returns

The results of the economic analysis are summarized below from details presented in Table SC 12.

| Sal-Coppice Forest System | Net worth (Rs.000, 12%, 30 yrs) | | | |
|---|---|---|---|---|
| | With JFM | Without JFM | Increment | % |
| Natural forest | 2,546 | 2,573 | -26 | -1 |
| Initial coppice | 762 | 0 | 762 | - |
| Plantation | 181 | 41 | 140 | 342 |
| Overall | | | | |
| - Excl. initial coppice | 2,728 | 2,614 | 114 | 4 |
| - Incl. initial coppice | 3,490 | 2,614 | 876 | 34 |
| - Excl. initial coppice per ha (130 ha) | 21 | 20 | 1 | 5 |
| - Incl. initial coppice per ha | 27 | 20 | 7 | 35 |

The data show that the introduction of JFM leads to a small increase (4%) in the overall value of the village managed forest. This excludes the value of the initial coppice at the start of the ten-year cycle. In practice, in many areas this has been delayed and the introduction of JFM brings forward the harvesting cycle and, therefore, this should be treated as an economic gain. The value of this gain depends on the length of the delay which would have occurred without JFM. Assuming a two-year delay on average, raises the incremental benefit as a result of JFM from 4% to 34%. Although the overall economic gains due to JFM are relatively smaller than in the Mixed Teak forest system, the value of the forest is still substantial, at Rs 27,000 per ha.

Wood and NTFPs contribute equally to the value of the forest before the introduction of JFM; however, with JFM, the contribution to total net worth, excluding the value of the initial coppice, falls from about 50% to 22% of total net worth. More surprisingly, there is an absolute fall in the value of NTFP of nearly 30%. Most of this is due to reduced tendu production as the canopy cover builds up under JFM as shown below.

| Sal Coppice Forest System (excluding initial coppice) | Net Worth (Rs.000, 12%, 30 yrs) | | | |
|---|---|---|---|---|
| | With JFM | Without JFM | Increment | % |
| Natural forest | 2,546 | 2,573 | - 27 | -1 |
| - Wood | 1,767 | 1,320 | 447 | 34 |
| - NTFP | 876 | 1,252 | 377 | - 30 |
| - costs | - 95 | | | |
| Plantation (net benefit) | 181 | 41 | 140 | 341 |
| - Wood | 365 | 0 | 365 | - |
| - NTFP | 10 | 41 | -31 | - 76 |
| - costs | - 195 | | | |
| Total | 2,728 | 2,614 | 114 | 4 |
| - Wood | 2,132 | 1,320 | 812 | 59 |
| - NTFP | 596 | 1,293 | - 697 | - 29 |

## Revenue Sharing

Revenue, defined as the revenue from timber and bamboo harvest yields only, is shared between the FD and FPC in the proportion 75% to 25%. All other outputs are assumed to belong to the community. The value over 30 years of the FPC share of harvest revenue amounts to Rs 844,000 excluding a share of the initial coppice. On an annualized basis, at 12%, into perpetuity, this is equivalent to Rs 70,000 per annum (or Rs 550 per household per year). The FD share amounts to Rs 1,203,000 (at 75% share of final harvest). This is equivalent to an annual flow of revenue of Rs 100,000 per year into perpetuity (approximately Rs 770 per ha per year). The figures are summarized below:

| Sal Coppice Forest System | Net Worth (Rs.000, 12%, 30 yrs) | | | |
|---|---|---|---|---|
| | With JFM | Without JFM | Increment | % |
| FD revenue | 1,203 | 376 | 827 | 220 |
| FPC revenue | 844 | 0 | 844 | - |

## Income and Distributional Impact

Although the analysis shows that the FPC gains revenue of Rs.844,000 from the revenue sharing agreement with the FD, there is a decrease of about 20% in the total FPC income, equal to a loss in net worth of Rs.7,500 per household. This is equivalent to an annual flow of income of Rs.625 per year into perpetuity at 12% discount rate, or 40 days paid labor per household.

The major stakeholders within the community have been identified in Chapter 3 and the value of returns to the various groups is presented below [10].

| Stakeholders | Net worth (Rs.000, 12%, 30 yrs) | | | |
|---|---|---|---|---|
| | With JFM | Without JFM | Increment | (%) |
| FPC (Revenue) | 844 | 0 | 844 | |
| Employment | 467 | 0 | 467 | |
| Headloaders | 0 | 1,365 | - 1,365 | |
| Livestock owners[11] | | | | |
| NTFP collectors | 2,131 | 2,928 | - 797 | -27 |
| Net in kind | 2,598 | 4,293 | - 1,695 | -39 |
| Total FPC income | 3,442 | 4,293 | -851 | -20 |
| Per HH | 31 | 38.5 | -7.5 | |

The major losers within the community, as result of JFM and protection, are headloaders, who are often from the poorest sub-group within the village. There is, therefore, a danger that without compensating activities, JFM arrangements will disadvantage households who depend on headloading for a livelihood. NTFP collectors are also losers under JFM. As was noted earlier, this result appears to be more surprising than for headloading. The reasons for this are as follows:-

- the main reason, accounting for 40% of the reduction in NTFP, is due to reduced tendu production from plantation areas;

---

[10] The analysis of the distribution of benefits within the community depends on the income effect of the transfer of assets and was, therefore, carried out using market prices based at the point of sale/purchase of inputs and outputs.

[11] Although the impact of JFM on grazing is generally recognized as important, this was not the case in the study villages, so data have not been included

- the second reason relates to mushroom production. Where annual hacking does not destroy the canopy cover and leaf litter is recycled, mushroom yields remain high. However, with regular harvesting and multi-shoot cutting every 3 or 4 years, the canopy is opened up and overall mushroom production falls.

## Management or Control Variables

The impact of different assumptions about the control variables identified earlier was analyzed. These variables include the institutional and regulatory framework, project interventions such as plantations, and FPC management practices related to offtake or hacking rates.

**The Institutional and Regulatory Framework**. The sharing arrangements between the FD and FPC differ from state to state. In West Bengal, there is pressure from FPCs for the Government to raise the FPC share - perhaps to 50%, the traditional share-cropping proportion, where inputs are provided by the landlord.

| Sharing Arrangements | | Net Worth (Rs.000, 12%, 30 yrs) | | | |
|---|---|---|---|---|---|
| | | With JFM | Without JFM | Increment | % |
| Present rate | FD (75%) | 1,203 | 376 | 827 | 220 |
| | FPC (25%) | 844 | 0 | 844 | - |
| Equal shares | FD (50%) | 770 | 376 | 394 | 105 |
| | FPC(50%) | 1,277 | 0 | 1,277 | - |
| Three quarters to FPC | FD (25%) | 337 | 376 | -39 | -10 |
| | FPC (75%) | 1,710 | 0 | 1,710 | - |

The switching value below which the FD incremental revenue becomes negative was estimated to be 27.5%, assuming that all intermediate yields go to the FPC.

**Plantations**. Most of the incremental benefits for the model village arise from planting. The ERR (30 years) for plantations was estimated to be 18% compared to 10% for natural forest operations and 15% for the village as a whole. These estimates do not include FD overhead costs. Increased areas under plantation are likely to have lower returns since the without-project yields will become progressively higher as new areas are selected.

# V. DISCUSSION

## General

The main objective of the study was to develop a better understanding of the incentives for various groups of forest users to participate in JFM. To achieve this objective, it was necessary to develop an appropriate analytical method. The method and its limitations are first discussed briefly below. In the Section B of this chapter the incentives for groups of forest users are summarized and discussed. However, the conclusions cannot be widely extrapolated as general findings applicable to JFM arrangements throughout India, as the ability of local communities to factor in specific values and perceptions into decision-making depends on the enabling environment in which decisions are made. Various factors influencing this enabling environment are discussed in the Section C of the chapter and implications for the future in Section D.

## Analytical Method

The analytical method developed during the course of the study provides a means of assessing both the overall returns to JFM, to both the FPC and the FD, and the distributional impact within the community. Although the analysis was based on field data, it is dependent on the assumptions used and it is recognized that three important assumed relationships require validation:

- CIP was used as a measure of forest growth, but the lack of data on annual growth in natural forest that is subject to hacking, meant that relationships between MAI, CAI and CIP had to be derived from standard yield tables;

- NTFP productivity was related to tree growth and canopy cover, but these relationships are complex and may be site-specific, so the assumptions used may be oversimplifications;

- Fodder availability was in only one case related to tree growth and canopy cover, using a simple formula.

Thus, although the method provides a systematic framework for analysis, it requires further testing and development in different ecological and social conditions.

## The Incentives for JFM

Joint Forest Management (JFM) was defined in Chapter 1 as "the sharing of products, responsibilities, control, and decision making authority over forest lands, between forest departments and local user groups, based on a formal agreement". The underlying assumption is that there is a convergence between the private incentives of forest users and the national objective of maintaining forest resources, and that people will protect the forest because they have a stake in the outputs. The analysis undertaken for this study suggests that this assumption is valid, at least for the representative villages in the two selected forest systems. However, although there appear to be overall benefits to JFM in economic terms, which are more secure in terms of revenue, some sub-groups within the FPCs may be adversely affected in the short-term. The extent to which these groups are protected depends on the internal dynamics and functioning of the FPC, as discussed in Section C.

**Benefits**

**Economic Returns**. The results of the analyses presented in Chapter 4 show that the net worth of the forest areas under FPC management increases with JFM. In the Mixed Teak Forest System, the increase is substantial, amounting to 58%, largely due to better management of the natural forest. In the Sal Coppice Forest System, the introduction of JFM provides a small increase of 4% in the overall value of the village managed forest, though if the initial coppice is included, the increase in value amounts to 34%.

The contribution of NTFPs to net worth of both forest systems is significant. Without JFM it amounts to 26% in the Mixed Teak Forest System, and 50% in the Sal Coppice Forest System. With JFM these proportions change, rising to 34% in the Mixed Teak Forest System and falling to 20% in the Sal Coppice Forest System, largely due to reduced tendu production, as the canopy cover builds under JFM. This illustrates the danger of making general assumptions about relationships between NTFP production and JFM.

**Revenue Sharing- FPC**. For the FPC, the introduction of JFM provides a substantial source of revenue from timber and bamboo yields, since before the introduction of JFM, the community did not receive any share of the revenue from harvest. The actual flow of funds to the FPC depends on the sharing arrangements, the size of the area handed over, and the number of members of the FPC. In the mixed teak forest system, the FPC revenue share from harvest amounts to Rs.52,000 per household. Overall, there is a threefold increase in the value of income flows into the FPC from Rs.33,000 per household to Rs.110,000 per household (equivalent respectively to Rs.2,730 and Rs.9,300 per annum into perpetuity at 12%). In the sal coppice forest system the FPC share of revenue from harvest amounts to Rs.440,000, equivalent to Rs.70,000 per annum into perpetuity at 12%, or Rs.550 per household per year. However, there are significant reductions in income for headloaders and NTFP collectors, amounting to the equivalent of 40 days paid labor per household. The gains in harvest revenue to the FPC in the mixed teak forest system are greater than in the sal coppice forest system, mainly because in Gujarat the FPC obtains a 50% share of revenue from harvest, as opposed to only 25% in West Bengal. In both cases studied, the gains are totally positive, but they will only be realized if the community is confident that protection will be effective. In addition, since much of the revenue is not realized immediately, incomes must be maintained from other sources, in particular, for the losers within the community.

**Revenue Sharing - FD**. Prior to the introduction of JFM, all revenue from the forest was collected by the FD. Since JFM introduces revenue-sharing arrangements, the analysis shows the expected gains to the FPC and apparent losses to the FD. However, the percentage realization of potential revenue without JFM is not known. In areas of heavy degradation it may be close to zero, whilst theft and smuggling, which is likely to be higher without JFM, will further reduce revenue. With

JFM, the community have an incentive not only to protect the forest but also to ensure that harvesting takes place on time and, therefore, to ensure revenue recovery. Consequently, the overall revenue benefits of JFM to the state are, in practice, likely to be substantial.

**Income and Distributional Impact**. The introduction of JFM shows overall economic benefits at the village level, which together with revenue sharing create clear incentives for the FPC to participate. However, within the community the incentives are not so clear cut. The analysis shows the impact of JFM on four groups of stakeholders in the communities: revenue earners, wage earners, fuelwood headloaders, livestock owners, and NTFP collectors. These groups are not independent of each other so the value of returns to individual groups does not reflect individual incomes. NTFP collectors gain substantially from the introduction of JFM in the mixed teak system, but lose in the sal coppice system, mainly because of reduced income from tendu leaf collection. In both systems, the main losers are fuelwood headloaders, who are often from the poorest sub-group within the village. Thus, the analysis indicates a transfer from poor sub-groups towards wealthier groups who can best afford to wait for the revenue from harvest. There is a danger, therefore, that the introduction of JFM arrangements will disadvantage the most vulnerable households. It is important, therefore, that compensatory activities be targeted at the poorest groups.

## The Enabling Environment

A key element of participatory forest management is that decisions are based on local understanding and can be modified in response to site specific experience. The ability of local communities to make decisions that take account of the values and perceptions of different stakeholders, depends on the enabling environment, that is on complex institutional and social conditions. Consequently, the realization of the benefits discussed above depends on the enabling environment in which JFM is implemented. In particular, it depends on:

- The internal dynamics and functioning of the FPC, in terms of membership and decision-making, which is critical to group understanding of biological relationships and economic values;

- The institutional and regulatory framework, which defines responsibilities and implementation arrangements, establishes sharing arrangements on the basis of a formal agreement, effectively a contractual arrangement, and sends signals which affect decision-making; and

- The regional economic context.

Because of the complex inter-relationships between these factors and site-specific ecological factors, it was not possible to take account of them in a quantified manner in the analysis. However, they are discussed below in the light of data collected from a number of villages in the two study areas and information contained in the literature.

## Internal Dynamics of the FPC

**Membership**.    Membership of the FPC is open to all households within the village.   The introduction of JFM has, in some cases, built on existing patterns of use, while, in other cases, JFM has introduced new stakeholders and altered existing, often informal, forest dependency relationships.   In small, homogeneous user communities the introduction of JFM can result in an increased sense of responsibility for the forest and, in general, an improved relationship between FD and community.   Other benefits include a sense of pride in the development of group rules and norms, and satisfaction at the legitimization of a share of benefits going to the community.   The introduction of JFM into larger, more heterogeneous communities may result in internal conflicts, with decision-making more likely to be influenced by existing power-relationships.   In particular, the effect of JFM is, often, to:-

- Introduce new stakeholders.  As a result of the revenue-sharing arrangements, a new and powerful class of stakeholder can gain rights within the forest.  Larger farmers, with little dependency on the forest who were previously largely uninterested in forest products, become more interested as a result of acquiring an explicit right to a share of forest benefits under JFM.

- Limit existing stakeholders.  Sub-groups, who prior to the introduction of JFM had been regulating their own use informally, are now subject to effective local policing under JFM arrangements.

The interests of these sub-groups do not overlap: for non-dependent sub-groups, benefits are maximized through timber harvesting; existing users, however, may be more interested in taking out NTFP and fuelwood, in kind, over the life of the forest.  The marginalization of some sub-groups may be reflected in a feeling of being disadvantaged due to lack of information about the agreements with the FD and the scope of JFM.

**Leadership**.  The leadership of the FPC tends to be in the hands of the better educated local elite, mostly from wealthier castes whose prime interest is a share of harvest revenue.  This group have the least to lose from effective protection and the most to gain from deferring benefits.  By definition, the local elite have better access to the FD and, therefore, are more able to ensure good information flows and working relationships with the FD.  However, the elite tend to be less dependent on the forests for their livelihoods and, as a result, decision-making tends to reflect the thinking of these revenue-minded groups within the FPC.  This results in marginalization of the poorer, usually more forest dependent, groups.  Further, there are existing socio-economic relationships between sub-groups within the FPC - for land, labor and credit - which need to be maintained within the village and which may make it difficult for indebted sub-groups to voice their opinions openly.

The Executive, or Management Committee tends to make decisions on behalf of the FPC, though this is subject to ratification.  The composition of this committee is, therefore, critical to ensure that the perspectives of all stakeholders are  reflected in decision-making.  However, in large groups, such as are common in West Bengal, it is not possible to represent all sub-groups on the committee. In both West Bengal and Gujurat there is an explicit role for the Panchayats in the FPC. In West Bengal the leadership of FPCs has tended to become politicized, with party workers and panchayat members becoming FPC leaders.

**Women's Participation.** In both the FPC and the Executive or Management Committee the participation of women is generally low, despite provision for women at all levels. There are two main reasons for this: (a) meetings are scheduled in the evenings to suit men, but at times when women tend to be cooking and, (b) when attending meetings, women feel marginalized, unlistened to and shy to talk. As result, there is a bias in favor of those forest products of interest primarily to men. Since it is women who are the main collectors of fuelwood and NTFPs this reduces the overall perceived benefits and eschews decision-making away from NTFPs despite the relatively high net worth of these products.

## The Institutional Framework

**Approach of the FD.** JFM differs from traditional rural development schemes, because it focuses primarily on the transfer of rights and responsibilities, and only secondarily on the transfer of resources. More usually, resources are transferred to rural areas and targeted rural groups through projects or schemes implemented by Government Departments, including the FD. Responsibility for the implementation of such schemes and for the maintenance of assets created remains with the Department, rather than with the community. The different emphasis of JFM requires a major change in the approach of the FD, in particular, requiring FD staff to release control of resources. However, for many, both within the FD and FPCs, JFM is regarded as just another channel for public sector investment, or employment generation, rather than as a means to protect forest assets on a sustained basis by the community. This perception reduces the incentive for FPCs to protect and manage the forest and, in practice, it is re-enforced by the control exercised by the FD over micro-planning and the formulation of by-laws. The FD remains in the role of giver; the community in the role of recipient.

**Group Formation.** The process of group formation determines the relationship between the FD and the FPC and the degree of independence the FPC claims. It is the FD that establishes and registers FPCs and assists in the preparation of a microplan for forest management and benefit-sharing. The process of group formation under JFM differs not only from state to state but from village to village. In Gujurat, NGOs were initially instrumental in negotiating and establishing JFM-type arrangements with the FD. In West Bengal, the pressure for JFM came from communities themselves, assisted by FD staff.
The legal status of FPCs also differs in different states. In West Bengal, the FPCs do not have independent legal status, but are attached to the FD for the purpose of JFM. In Gujurat, FPCs are required to become independent co-operatives capable of utilizing funds from various schemes and programs. This reflects the role of non-FD organizations, especially NGOs, in the initial formation of FPCs.

**Planning**. FD operations are based on Forest Working Plans (FWPs), that contain baseline data and management prescriptions. However, FWPs often provide only a poor reflection of conditions on the ground, due to delays in updating. In addition, FWPs are not oriented to multiple objective management and are seldom based on consultation with local forest users. By contrast, an essential element in the introduction of JFM is the preparation of a village microplan for the area controlled by the FPC. This is based on participatory consultations between the community and FD, and aims to reflect the priorities and needs of FPC members, within the framework of sound sylvicultural practice. Consequently, FWP prescriptions are frequently in conflict with the agreed microplans. This has implications not only for choice of species and forest management, but for harvesting, as discussed in Para.

**Allocation of Forest Land**. In many cases, FPC boundaries have not been formally demarcated as natural, administrative and customary boundaries do not overlap. In practice, under existing, customary use, different boundaries apply to different products, eg grazing and fuelwood. Boundary disputes between neighboring FPCs are likely to increase as harvesting approaches.

The forest areas allocated to FPCs are either plantations or degraded natural forest with a canopy cover of less than 40%. In some cases the allocated area is less than the area that was customarily associated with a particular FPC and this has created resentment within the FPC and resulted in failure. In parts of Rajpipla, where forest land has a canopy cover greater than 40%, the Government Regulation does not permit JFM arrangements and it has not been possible to hand over the forest area to an FPC, even where the community was willing to participate. Instead the FD have constituted FPCs, charged with forest protection, and routed FD protection costs through the committee. There is debate, both within and outside the FD, as to whether JFM sharing arrangements might be possible and whether, in areas of good forest, the share to the community should be less than for other types of FPC.

**Protection**. Protection of the forest is seen as the most important activity in JFM, by both the FPC and FD. In most functioning FPCs protection involves patrols by groups of members on a regular cycle of one week or ten days. Initially protection activities are intense and difficult until an accommodation is reached with neighboring communities. Once the system and boundaries are established, the intensity of activity reduces. In West Bengal, the FPC needs to establish its claim ("atta") to the area.

FPC protection has resulted in an increase in the number of offenders being caught and, increasingly, FPCs are fining offenders. However, the legality of FPCs retaining the fine is uncertain, so FPCs are generally encouraged to hand the fine over to the FD. However, in some cases fines are included in FPC accounts and, in one case, the FPC even uses part of the fine to reward the catcher. In some tribal groups, fining is not socially acceptable so persuasion or social sanctions are used. Most VFCs want to handle local offenders themselves, but look to the FD for strong action against offenders who can resist local pressure, either violently or through connections. This results in an increase in the FD workload.

Previous protection arrangements for schemes under which the FD hired and paid for watchers are now in conflict with the new JFM arrangements, with watchers resentful that their role has been transferred to FPCs. At the same time, FPCs are sometimes unsure of their responsibilities for protection while the old arrangements are still in place.

**Planting and Support Activities**. JFM is often associated with a planting program, organized by the FD, as well as protection of specified areas of forest. The planting program provides employment opportunities for local labor and, for many, this may be the most immediate incentive for JFM. The association of JFM with employment generation in new plantations means that it is difficult to determine other perceived benefits of JFM.

Apart from plantation work, JFM is often linked to other support activities, either to create community assets or for income generation. These schemes are funded from other sources, such as World Food Program funds. FD staff regard these activities as a means of (a) encouraging participation in JFM, and (b) taking the pressure off the forests by increasing local incomes. In general, it appears that FPC members do consider these schemes as an incentive to participate in JFM. The extent to which the whole community participates in the choice of activity varies and, in practice, there is a tendency for richer groups to gain most of the long term benefits from asset creation, since structures like checkdams provide irrigation to landowners. Arrangements for repair and maintenance of the assets, for example biogas plants, are often ill defined, which limits the duration of any benefit. The desirability of these support schemes is widely debated within FDs, as the benefits in promoting JFM must be set against the risk of promoting a culture of patronage or dependence in the relationship between FD and FPC. Greater involvement of the FPC in decision-making, and increased responsibility by the FPC for implementation of the schemes, would help to minimize these problems.

**Harvesting Arrangements**. These may be as important to the FPC as the share of the produce. In West Bengal, where harvesting under JFM is about to start, harvesting is in accordance with the prescriptions of the Working Plan. In Gujarat, the FPC applies to the DFO to harvest an area using an application form which, itself, constitutes a working scheme. However, the Working Plan area may include many individual FPC microplans, so that FPC members' expectations are in conflict with Working Plan prescriptions.

All harvesting, in reserved forest areas is currently carried out either by the FDC, FLCs or through industrial units by contract or agreement with the FD. The FPC's share of the revenue is net of harvesting and marketing costs and, therefore, harvesting arrangements directly affect the return to the FPC. FPCs argue that harvesting costs are higher than normal as result of the monopoly position of the Forest Development Corporation (FDC) or Forest Labor Cooperatives (FLCs). For example, a comparison of prices for two adjacent fields of Eucalyptus poles in West Bengal showed that after deductions for harvesting costs, each pole on private land would fetch Rs.8, while on FD land, harvested by FDC, the FPC would receive only Rs.1 for each pole.

The arrangements whereby the FDC harvests timber on behalf of the FD, even in forests under JFM, creates difficulties for the FDC as they are unable to decide the timing of the  harvest and, thereby, to maximize the sale price, or to control their own harvesting costs.
FLCs suffer similar problems as rates are fixed by the FD with the FLC federation on behalf of its members.

Harvesting contracts with industries are often for a fixed period and for given quantities.  In many cases, the contracts are complicated, often involving the contractee in harvesting, for example, bamboo and then returning unwanted portions, eg poles, back to the FD.  Previously, many of these contracts were fixed at low subsidized prices but, in general, contracts are now linked to average auction prices.  However, more importantly, the contracts are agreed between the FD and the concerned enterprise and negotiated at headquarters level.  At the local level, FD staff have to meet targets in order to honor these contracts.  The FPCs are aware of these contracts and, in many cases, perceive them to be against their own interests.

The way in which communities value benefits may also affect decisions on harvesting and marketing.  Forest  benefits consist of many products which mature over different time periods and which can be realized in different ways at different rates.  Time preference rates differ within the community, so it is difficult to value future outputs and determine the benefits of alternative actions, creating uncertainty about future benefits.  There may also be  uncertainty within the FPC about the form in which benefits will be realized: cash or kind, individual or community, which fuels mistrust between sub-groups within the FPC.

**Marketing Arrangements**.  The FDC also has responsibility for marketing timber and poles on behalf of the FD, even under JFM arrangements.  The FPC has, therefore, no control over the prices obtained, and hence of the value of their share in the final harvest.  Similarly, monopoly control over the marketing of some economic NTFP has been assigned to either to tribal cooperatives, known as Large -sized Multi-Purpose Cooperative Societies (LAMPS), or FDCs. Membership of a LAMPS is restricted to tribals, resulting in conflict with the FPC, since control over NTFP collection and sale is claimed by two only partially overlapping groups.

### The Regional Context

The study did not analyze the economic effect of the gradual introduction of JFM within a region as a whole.  However, the positive economic impact of JFM is reflected in the increase in the net worth of the forest shown by the analysis, and the environmental benefits, both in situ and downstream, of an increase in the area of forest cover [12].  A rapid spread of JFM, resulting in large scale partitioning and protection of the forest is, in the short term, likely to reduce supply, but may increase prices and the incentives for protection.  JFM could also have a negative regional economic impact as a result of increased exploitation of adjacent forest areas not

---

[12]     In Madhya Pradesh the environmental benefits resulting from a state forestry project were estimated at Rs.376 (About US$12) per hectare, representing the difference between the economic returns to the project and an OCC of 12%.

protected through JFM. This might be a short-term impact, if JFM arrangements spread rapidly. JFM could, however, also lead to more rapid degradation of relatively undisturbed forests with a canopy cover exceeding 40%, in which JFM is not presently permitted. The balance between these positive and negative impacts are dependent on a number of factors, such as the supply of wood, particularly fuelwood from community plantations, and employment opportunities associated with logging and plantation activities. The regional economic impact of JFM is also strongly linked to many factors which lie outside the forest sector, of which increased employment opportunities outside forest areas is one of the most important.

## Implications for the Future

**Community Dynamics.** The results of the analysis shows that although the introduction of JFM results in positive economic benefits to the community, there are some sub-groups who may be adversely affected. As these groups tend to be the most disadvantaged, the introduction of JFM must be associated with special measures to compensate these groups. In particular, these sub-groups must be adequately represented on the executive committee of the FPC. Members of the sub-groups should get preferential access to employment opportunities in the forest sector. Village assets created through programs to support JFM must be based on the priorities of the disadvantaged, rather than those of the elite. Without such special measures the disadvantaged will continue to exploit, rather than protect the forest resources, or become further marginalized.

**The Institutional Framework.** The study emphasizes the widely recognized need for changes in the traditional approach of the FD, if JFM is to be successfully implemented. In addition, the study has shown that current FWP prescriptions are often in conflict with the village microplans, superimposing targets, and preventing FPC participation in decisions concerning harvest. There is, therefore, an urgent need to suspend FWP prescriptions in areas of JFM. Existing commitments of the FD, particularly in relation to harvesting, may reduce the returns to the FPC. Thus, institutional arrangements and policies must reinforce the philosophy underlying JFM, sending clear signals concerning incentives to decision-makers within the both the FD and the FPC.

**NTFPs.** The study indicates the importance of NTFPs to the net worth of both the forest systems studied. It also illustrates the complexity of the relationships between NTFP production and JFM. This means that implementation of JFM programs must be based on site specific considerations and community priorities. The study also highlights the equity implications of changes in NTFP production.

**Further Work.** The results of the study depend to a large extent on the relationships and data included in the models. The data was collected in the field, but in a very limited number of sites. The following studies are, therefore required:

- Application of the analytical method in a variety of ecological and social conditions to test the extent to which the conclusions can be extrapolated;

- Estimation of the annual increment in degraded natural forest, less than 10 years old, under different biotic pressures, in order to establish the relationships between MAI and CAI and calibrate the analytical model;

- Measurement of NTFP productivity in relation to forest degradation in a variety of ecological conditions;

- Assessment of fodder resources, both herbaceous and tree fodder, derived from forest of different levels of degradation;

- Investigation of the internal dynamics of FPCs, with particular reference to the representation of disadvantaged groups.

## Implications for Bank Lending

The results of the analysis in the two areas studied confirm that the introduction of JFM has positive economic benefits and financial benefits to participating communities. Although the results can only be extrapolated with caution, they suggest that the conclusions of earlier analyses undertaken during appraisal of Bank-supported forestry projects have been correct and that Bank financing for JFM programs has been appropriate. However, the conclusions of the study with regard to the uneven distribution of benefits within the community, point to the need for future project to include specific measures to safeguard the interests of those who may lose from the introduction of JFM, especially as they are often the poorest and most vulnerable in the community. These measures may include efforts to ensure that the losers get preferential access to any project-financed employment, or that the project includes measures to develop alternative sources of income for these groups.

The study is also relevant to Bank lending in other sectors in which community management of resources is being introduced, for example, community management of irrigation schemes. The economic and financial implications clearly cannot be extrapolated, but the analytical approach, and in particular, the need to consider the distributional impact of schemes that appear to benefit communities as a whole, do have relevance to other sectors.

**Annex 1**

**Preliminary Report on Study of Incentives for Joint Forest Management**

Main Report

      Study of the Incentives for Joint Forest Mangement

Working Papers:

    1.     Background to JFM in India

    2.     Overview of JFM Arrangements

    3.     Field Study Report:  Bankura

    4.     Field Study Report:     Rajpipla

Team Members:

| | |
|---|---|
| Economics | D. Shields (Team Leader) |
| | K. Shah |
| Social Science | Ms. M. Raju |
| | Ms. A. Benninger |
| | Ms. E. Bazellgette |
| Forestry | M.H. Swaminath |
| | A. V. R. K. Rao |

Femconsult,
Koninginnegracht 53, 2514 AE,
The Hague, The Netherlands

## Annex 2

## Forest Productivity

**Measures of Productivity**. Conventionally, productivity is measured using Mean Annual Increment (MAI) over a full rotation. However, this measure does not adequately reflect annual offtake and current Annual Increment (CAI) provides a better measure of yield to reflect offtake during the rotation. Annual CAI was derived from standard yield tables. Actual CAI depends on Growing Stock (GS). In this study, actual GS was calculated from field survey data on number of stems, basal area, girth, height, together with standard form factors. This provided an estimate of growing stock at a point of time. Information on the amount of growing stock which has already been harvested and the rate of future growth under protection was estimated by comparing similiar sites under different levels of protection. Additional data from plantations and working plans gave the relationship between age and growing stock for different site qualities. The problem was to apply this information to calculate CAI for natural forest, with trees of different species and a wide range of ages, especially at early stages of tree growth where, without protection, hacking takes place.

**Method of Estimation**. The Current Increment Percent (CIP) is the relation of the increment during a given year, to the volume at the beginning of the year. The CIP, therefore, defines the CAI as a percentage of current GS. The CIP was calculated from standard yield tables on the basis of total yield. The relationship between CIP and GS was estimated using an inverse, parabolic reltionship for the mixed teak forest system and for the sal coppice forest system as shown below. A number of other functional forms were tried and rejected.

**Mixed Teak Forest System**. A number of different species occur within this system, but the relationship was estimated using teak because data for other species are not available. This can be justified because teak is invasive in nature and likely to dominate over time.

The estimate is based on the data shown in Table 1 below. The relationship that explains 99% of variation was estimated to be:

$$CIP = 8.43 + 532/GS \qquad R^2 = 0.99$$

Table 1.  Teak Productivity

| Year | GS Final | GS Total | CAI Total | CIP % | 1/GS Final |
|------|----------|----------|-----------|-------|------------|
| 1    | 5.00     | 5.00     | 5.000     | 100.00 | 0.200     |
| 5    | 15.00    | 15.00    | 3.000     | 20.00  | 0.067     |
| 10   | 26.94    | 26.94    | 2.694     | 10.00  | 0.037     |
| 15   | 34.99    | 42.33    | 3.078     | 7.27   | 0.029     |
| 20   | 41.28    | 55.63    | 2.660     | 4.78   | 0.024     |
| 25   | 48.48    | 66.12    | 2.098     | 3.17   | 0.021     |
| 30   | 50.38    | 76.62    | 2.100     | 2.74   | 0.020     |
| 35   | 53.18    | 84.32    | 1.540     | 1.83   | 0.019     |
| 40   | 58.08    | 92.00    | 1.536     | 1.67   | 0.017     |
| 45   | 61.23    | 100.06   | 1.612     | 1.61   | 0.016     |
| 50   | 65.78    | 107.76   | 1.540     | 1.43   | 0.015     |

Source:  Standard Yield tables (Site Quality IV)

GS (Final) = Standing Growing Stock

GS (Total) = Standing Growing Stock + Intermediate yileds

CIP% = 8.43 + 532 (1/GS)

**Hyperbolic Form**

Regression Output

| | |
|---|---|
| Constant | -8.42916 |
| Std. Err of Y Estimate | 2.633276 |
| R Squared | 0.9926 |
| No. of Observations | 11 |
| Degrees of Freedom | 9 |
| X Coefficients | 532.0975 |
| Std Err of Coef. | 15.31379 |

**Sal Coppice Forest System**. The estimate is based on the data shown in Table 2 below. The relationship was:

$$CIP = 7.44 + 0.236/GS \qquad R^2 = 0.70$$

Table 2. Sal Productivity

| Year | Stem Volume (SV) (cm$^3$) | CAI | CAI/SV % | 1/SV |
|------|------|------|------|------|
| 1 | | | | |
| 2 | | | | |
| 3 | 4 | 0.004 | 100 | 263.2 |
| 4 | 5 | 0.001 | 24 | 200.0 |
| 5 | 7 | 0.002 | 29 | 142.9 |
| 6 | 9 | 0.002 | 22 | 111.1 |
| 7 | 12 | 0.003 | 25 | 83.3 |
| 8 | 17 | 0.005 | 29 | 59.5 |
| 9 | 22 | 0.005 | 24 | 45.0 |
| 10 | 30 | 0.007 | 25 | 33.9 |
| 11 | 38 | 0.008 | 21 | 26.7 |
| 12 | 45 | 0.008 | 17 | 22.0 |
| 13 | 55 | 0.009 | 17 | 18.3 |
| 14 | 62 | 0.007 | 12 | 16.1 |
| 15 | 71 | 0.009 | 12 | 14.1 |
| 16 | 81 | 0.010 | 12 | 12.3 |
| 17 | 90 | 0.009 | 10 | 11.1 |
| 18 | 100 | 0.010 | 10 | 10.0 |
| 19 | 109 | 0.009 | 8 | 9.2 |
| 20 | 118 | 0.009 | 8 | 8.5 |
| 21 | 126 | 0.009 | 7 | 7.9 |
| 22 | 135 | 0.009 | 7 | 7.4 |
| 23 | 143 | 0.008 | 6 | 6.9 |
| 24 | 153 | 0.010 | 6 | 6.5 |
| 25 | 161 | 0.008 | 5 | 6.2 |

CAI Rate = a + b/GS

Regression Output

| | |
|------|------|
| Constant | 7.442782 |
| Std. Err of Y Est | 10.91822 |
| R Squared | 0.697414 |
| No. of Observations | 23 |
| Degrees of Freedom | 21 |
| | |
| X Coefficients | 0.235918 |
| Std Err of Coef. | 0.03391 |

**Annex 3**

**The Mixed Teak Forest System. Gujarat**

**Description of the Study Area**

## General

Rajpipla Forest Division, located in the Baruch District of Gujarat, was selected as the field study area. Two villages, Dhanor and Magardev, in Fulsar Forest Range, were chosen as representative of villages in which JFM arrangements are in place. A number of other villages, Koyalivav, Gajargota, Garhi, Motia, Soliya, Khedipada, and Zarnavadi, were also selected to provide supporting information, or to represent conditions that differ slightly from the two main villages. Mean Annual Rainfall in the Baruch District is about 1,500 mm, falling during the monsoon season from June to October. April and May are the hottest months with temperatures reaching $45^0$C, with milder temperatures during the winter from October to February.

## Forest Cover

Most of the forests in the Rajpipla East Division are classified as slightly moist teak forests, though they occur in a transition zone to the dry teak forests of eastern-central Gujarat. Teak is the predominant species in both the types. There are better bamboo areas in the slightly moist teak forests, while the dry teak forests are known for greater availability of grasses and other non wood forest products (NWFPs). These include "veedies" or grass production areas in open forests, produce from which are auctioned by the FD. Marketing of forest produce has been common for a long period of time. Prior to the involvement of the British, timber harvest was organized by merchants from the plains, but this had little impact on the forests until the mid-nineteenth century, when demand for teak for ship building increased. As part of a princely state, the forests were severely exploited and this, combined to increased pressure of local populations on the forest has lead to degradation, often severe. About 20% of the forests in the area have a canopy density of 0.4 or above, and about 40% is either barren or highly degraded, particularly on the upper hill slopes.

## Floristic Composition

In the slightly moist teak forests, of the southern tropical moist deciduous forest sub-type in Champion and Seth's classification, teak is the predominant species, associated with sadad shisham, hed, kakad, bio, tanach, khair, hino, bondaro, kumbhi, behda, garmalo, kevlu, mad-mahuda and nagvel. Shrubs and herbs include madhuri, phuvadia, zipto, and jungli-bendi. Dry teak forests also occur in the area, usually on shallow soils on undulating hills. Teak is again the predominant species, but there are differences in the number and type of associated species.

## Current Forest Land Use

**Natural Forests**. As a matter of policy, the Forest Department has stopped all green felling in the natural forest areas, with the exception of khair trees to meet the commitment of the FD to katha factories. The FD supplies about 2 million bamboos at concessional rates to tribal peoples and has an agreement with the Central Paper Mill to provide bamboo, also at concessional rates. Fuelwood collection has a major impact on the forests, as many headloaders risk punishment by the Forest Guards to cut green timber and rootstock, as well as the dead and fallen wood that they are permitted to extract.

**NWFP**. The Forest Development Corporation has monopoly rights over the two main products, tendu leaf and mahua products. Medicinal plants, tubers and mushrooms are collected for subsistence or sold locally.

- **Tendu Leaf**. Leaves are collected and sold to traders, who compete for buying rights in specific areas. Collection provides a return of about Rs.60/ha to the collector. However, the quality of leaves is low and returns decline rapidly during the later part of the collection period, so collection is much less remunerative than in the past.

- **Mahua Products**. Even when they grow on common land, responsibility for protection and tending of specific mahua trees, and the rights of harvest from these, belong to particular households. These rights and responsibilities are recognized by all villagers. Flowers and seeds are collected for food, for making liquor, and for pressing oil. Higher yields are reported from trees in less degraded areas, and areas close to villages.

- **Medicinal Plants**. A government-run Ayurvedic Collection Center has been established in Rajpipla and has introduced a more systematic and controlled collection of medicinal plants to improve quality. The main plants are harda, behda, amla, katu, arni, and bilwa. In addition, there is a growing demand for safad museli, used in several ayurvedic tonics. Collection, drying and processing the plants is highly labor intensive, which reduces incentives for collection.

- **NWFP for Consumption**. A wide variety of tubers and leaves, fruits, vegetables, wild fruits, such as ber and beela, and wild varieties of beans and gourds have been collected traditionally.

**Grazing**. Forests provide the main grazing resource for livestock in the area throughout the year, with the exception of a short period of time after harvest, when livestock are permitted to graze on crop residues. Grass from the forests is also used for stall feeding, though this is generally restricted to draft animals during the agricultural season. Traditionally, livestock were herded in the forests by graziers who were paid in kind by the owners. This system is breaking down as more of the graziers migrate to towns or sugarcane plantations for daily paid labor. In addition to the grass required for their own animals, many grass collectors sell grass. Sale of grass is an increasingly important source of revenue for FPCs, with surplus grass being auctioned to bulk users, such as a nearby dairy cooperative. Estimated production of grass is about 1,700 kg/ha in the first year of plantation, falling to about 700 kg/ha in the fifth year.

## Socio-Economic Characteristics

**Population**. The population in most villages adjoining forest areas is almost entirely composed of tribal peoples. The majority are Bhils, of which there are two groups, the Vasavas and the Tadvis. Vasavas have their own dialect and claim to be a settled branch of the Bhils as they have settled in permanent hamlets of the area. Tadvis are less numerous in the area and speak Gujerati. There is also a small population of Kotwalias. In the two villages chosen for detailed study, all the people were tribals, mainly Vasavas.

**Agriculture**. Rainfed agriculture is the main occupation of people in the area. The main crops are paddy, jowar, tuvar, and coarse grains such as kodra, mor and banti. Some cotton is cultivated. Livestock are owned by many households and some villagers work as herdsmen. Wage labor plays an increasingly important role in most households, either as agricultural laborers, or after the agricultural season, as workers in neighboring towns. It is usually only the men who migrate, but there is an increasing trend for whole families to move to towns between November and March.

## Dependence on Forests.

The major dependence on the forests in the villages studied is for fuelwood and fodder. Fuelwood needs of between 400 - 800 kg per family per year are met from collection of dry wood, one cut-back operation and some green wood felled illegally. Forest sweepings and cattle dung only fill part of the fuel needs of the home, so a large proportion of families resort to headloading. Headloaders also sell fuelwood in nearby towns, despite low returns in terms of time and effort, and depressed prices due to the illegal nature of much of the collection. Small timber and bamboo for house construction, and timber for agricultural implements are all derived from the forest.

Dependence of communities on the forest for subsistence products has declined, due mainly to increasing scarcity, resulting from forest degradation. However, tubers, leaves, mushrooms and vegetables are all consumed by villagers, particularly those close to areas of less degraded forests. Villagers report that their own use of many of these products and of medicinal plants, has decreased as markets for the products have developed.

In Magardev, a major conflict with the FD began when an area of land being cultivated by the villagers was converted into a forest area, during the Forest Settlement process during the 1970s. The FD established plantations on the land as the villagers were unable to sustain the expenses of court action. Villagers persisted in cultivating these areas and had to pay fines.

Subsequently, some of the land has been returned to the villagers as the Government of Gujerat has regularized encroachment that took place prior to 1980. The resolutions for the regularization of encroached forest lands have served as a disincentive for protection of the forests, as they are seen in some communities as a means of procuring agricultural land for the landless. There is clear evidence of encroachment in both the villages studied in detail, either by residents of the villages, or by residents of neighboring villages.

**Community Perceptions of Incentives for JFM**

Participation in JFM requires some modification of a number of practices related to use of the forest resources, which may have short-term adverse impacts on villagers. The primary motivating factor appears to have been employment generation through the activities taken up in the area as a result of JFM. Plantation work is, therefore, often the starting point for successful JFM. However, NGOs in the area report that JFM can also be accepted without plantations. Support activities are also perceived as a major incentive, partly a source of employment, but also because of the benefits to agriculture or village society as a whole. It is also recognized that the supply of grass for cattle fodder increases substantially with JFM. A summary of perceived incentives and disincentives to JFM is presented in the following table.

## Villagers' Perceptions of Incentives and Disincentives Relating to JFM

| Protection-Related Activity | Benefits | Disincentives | | Comments |
|---|---|---|---|---|
| | | Observations | Coping Mechanisms | |
| Providing watchmen | Potential source of income in villages where FD/FPC decides to pay wages (usually unpaid) | * Where FD paying own watchmen, FPC members provide the protection while others are paid for it<br>* In villages with high seasonal migration for wage labor, human resources limited | FPC hires watchmen for season | Conflict where duel patrol/ protection systems exist reduces villager commitment to protection and causes disunity in the community, undermining the FPC |
| Protection of plantations: | Increased yield of grasses shared between FPC members, providing several months' supply from a few days work | Engagement to protect specific plots discourages recourse to plantations for fuelwood and fodder collection (illicit) as well as for grazing | Go greater distances to unprotected forest land or forests of neighboring villages | In general avoidance of grazing in plantation areas is already an accepted norm;<br>FD say plantation protection is important as an employment generation incentive for FPC, so make this a starting point, but payment is in kind, not in much needed cash. |
| Employment:<br>- in planting of new stock | | | | GR state payment in kind; where cash generated (e.g in sale of harvest) some FD officers believe cash should be invested in community development activities not shared between members |

| Protection-Related Activity | Benefits | Disincentives | | Comments |
|---|---|---|---|---|
| | | Observations | Coping Mechanisms | |
| Employment - in cleaning operations | Grasses and coppicing received help reduce need to depend on forest for fodder and domestic fuel needs; when there is surplus it can be sold to generate cash | | | |
| - in harvesting | * Will receive defined share of harvest (in kind) * Expect to be able to sell own shares through open auctions etc. to generate cash returns | * Experience to date with bamboo, sharing mechanisms not defined clearly, especially for marketing rights and systems; CPM removes harvest. * Previous practice for timber: collection from depots and provide transport back to village viewed as onerous. | One GVM staked claim to share of recent bamboo harvest and kept it in village. FD/CPM/GVM still debating situation. | * Not yet implemented for timber. * Very limited implementation for bamboo so far. Need to define basic future mechanisms and reconcile prior agreements with JFM commitments. * Some villages unclear as to share. * Differences of opinion within FD as to (a) role in marketing, and (b) acceptability of revenue from sales being shared directly between villagers. |

| Protection-Related Activity | Benefits | Disincentives | | Comments |
|---|---|---|---|---|
| | | Observations | Coping Mechanisms | |
| Protection of standing forest | Receive share of produce from cleaning operations; Have the right to forest sweepings for fodder and fuelwood | Forgo (illicit) green timber used to supplement fuel and fodder needs and as supply for head-loading to generate some income  Increased pressure on unprotected land due to greater demands (including from neighboring villages with FPCs)  In villages with good forest areas, increased pressure from grazing of livestock from neighboring villages with poorer grazing | Go greater distances to neighboring forests. | Women suggested that a special plantation plot for fuelwood and fodder supply would overcome many problems arising from fuelwood and fodder needs |
| Support activities by FD | Community benefits from activities which it might not implement otherwise | VS have limited choice of activities, proposed by FD, rather than setting own priorities | GVM, with NGO help, play greater role in decision-making | FD see employment generation as major incentive to FPCs; villagers value employment (especially when it brings cash wages) but also place a high value on impact on agricultural income |
| Water management (irrigation, gully plugging, check dams, bunding, talavdi repair/construction etc. | Increased productivity of agricultural lands, leading to increased income from land (enabling vegetable production and introduction of second crop); | | | Villagers perceive this as a major incentive |

| Protection-Related Activity | Benefits | Disincentives | | Comments |
|---|---|---|---|---|
| | | Observations | Coping Mechanisms | |
| Promotion of gobar gas plants | Reduced needs (time and quantity) for fuelwood collection | * Lack of skills in repair and maintenance within villages results in high proportion of biogas stoves being unusable; | | Access to repair & maintenance skills needed within village; High level of subsidy viewed by some NGOs as disincentive to necessary commitment to full and successful use; Only meets full cooking fuel needs for small households; |
| Provision of fish seed for talavdi | Potential source of income for community from fishery activities | | | |
| Fencing | Saves having to purchase/use allocation of bamboos in order to protect fields and plots from grazing damage; | | | Potential introduction of live fencing which would also serve as supplementary source of fuelwood and fodder closer to the village |

**Annex 4**

**The Sal Coppice Forest System, West Bengal**

**Description of the Study Area**

## General

Taldangra Forest Range, located in the northern part of Bankura (South) Forest Division in West Bengal, was selected for the purpose of the study. Within this area, three villages in two Forest Protection Committees (FPCs), at Fakirdanga and Shyampur, were selected for more intensive study and a number of other villages selected to provide supporting information. The climate in the region is sub-tropical with rainfall well distributed during the monsoon months, from mid June to September. Mean Annual Rainfall is 1320 mm. The mean maximum monthly temperature ranges from $38^{o}C$ in May to $12^{o}C$ in November. Sal forests in the area are usually associated with highly unsaturated red soils.

## Forest Cover

The original flora, according to Working Plan documents, was characterized by a fine coppice sal forest over an extensive area, but have been subject to excessive exploitation and severe biotic pressure from local communities. The Zamindari system during colonial times appears to have been the starting point for forest degradation. In order to meet the taxes levied on the Zamindars they exploited the forests without any rest period for regeneration. Over the years, rotations have been reduced from 70 to 80 years to a mere 5 years. About 3% of the forests have been worked with a rotation of 5 years, 70 % of 6 years, 2 % of 8 years, 10% of 8 to 10 years, 9 % of 10 years and 6 % of 10 to 14 years. In 1945 legislation was passed making it compulsory for the Zamindars to prepare working plans, approved by the FD, for the management of forests. However, the increase in population at the time of partition meant that the demand for forest resources increased rapidly, resulting in over-exploitation. Past FD extraction policy has also contributed to the degradation of the forests.

At present the forest cover is limited to uplands, hill slopes and ridges while the low lying areas and gentle slopes have been brought under cultivation. In the Bankura district, forest degradation is greatest towards the north-west and least in the south-east region. The northern and north-western parts, in particular, are devoid of any vegetation due to severe biotic interference and the dependence of the community on the forest resources.

## Floristic Composition

The forests in Bankura District are classified as Dry Peninsular Sal forests, according to Champion and Seth's classification of forest types in India. According to working plan descriptions, four types of forest formation are present:

- sal forests of coppice origin,
- sal mixed forests,
- khair formation forests,
- riverain forests.

In general, sal is the predominant species. The other main associated species are *Terminalia tomentosa, T. bellerica, Pterocarpus marsupium, Diospyros melanoxylon, Butea monosperma, Cleistosanthus collinus, Ougenia, Anogeissus latifolia, Lagerstroemia parviflora, Sterculia urens, Syzygium cuminii, Salmalia malabaricum, Madhuca latifolia, T chebula Phyllanthus embelica , Soymida fabrifuga, Terminalia arjuna Albizzia spp, Sapindus spp, Holoptelia integrifoilia.* The lower middle storey consists of *Holarrhena antidysentrica, Gardenia, Zizyphus, Flucartia, Randia* and *Ixora, Capparis* species. The ground flora, which is characterized by a lack of grass species, consists mainly of *Indigofera spp, Evelvulas, spermacoce, Barlaria spp, jatropa, Hemidesmus indica, ananthmul, Andrographis.*

## Current Forest Land Use

**Coppice Rotation**. The Forest Department normally manages sal forests on a 10 year rotation. Growth data show that on quality II and III sites, this results in about 68% of stems of less than 3" in diameter, 29% in the 4" diameter class, and 3-4% in the >5" class. Prices for 4" class poles are 300% higher than 3" poles and 5" poles a further 125% higher. The data indicate that it normally takes 3 to 4 years to gain one inch diameter growth in quality II and III sites, so extending the rotation by 5 years could provide more than 70% of poles in higher diameter classes, thereby maximizing returns.

Where protection measures are not enforced strictly, the sal forests are coppiced annually or in alternative years to meet fuelwood demand. This practice has caused considerable damage to the sal root stock as the coppiced shoots do not have sufficient time to establish as independent plants, which normally takes 3 to 4 years. The continuous and repeated hacking of stumps, and sometimes uprooting of the drying stumps has resulted in a gradual reduction in sal forest areas.

**NWFP**. The main products include kendu leaf, sal leaves (salpatta), sal seed, mahua seeds, mahua flowers, medicinal plants, mushrooms, and a number of other fruits and grasses:

- **Kendu Leaf**. At present, the greatest density of kendu plants is found in degraded forests (canopy <0.2) and other degraded open patches, although the plants are found in forests with a greater canopy cover. In general, it was found that the density of tendu varied between 3,000 and 1,500 plants/ha. Present sal forest management on a 10 year rotation may reduce the kendu productivity considerably.

- **Sal Leaves**. Collection of suitable sal leaves is greatest from younger trees, and the practice of MS cutting during the fifth and seventh years reduces production of sal leaves due to a reduction in the number of shoots.

- **Sal Seed**. There has been a gradual reduction in the yield of sal seeds due to degradation. Sal seed is collected in April and May. Production of seed appears to start in the third year and reaches its peak at 8 to 15 years age. The 10 year rotation practiced may, therefore, reduce the yield of sal seeds.

- **Mahua Products**. Mahua flowers are used both as a vegetable and for making liquor. The fruit is eaten as a vegetable or the pulp dried and burnt as an insecticide. The seeds are extracted and crushed to produce an edible oil. All these products are used for subsistence and sold. Particular households usually have rights to specific mahua trees.

- **Mushroom**. Production in the sal forests and plantation areas appears to be declining, due mainly to a reduction in forest floor biomass resulting from leaf litter collection.

- **Medicinal Plants**. A range of plants have been used for medicinal purposes but only kalmegh, anantmol, satmol,and kurchi are sold widely. The market is controlled by small traders and there are no data on the quantities sold.

**Leaf Collection**. Improved protection measures by both the FPC and Forest Department has meant that many people have resorted to collecting leaves for fuel by sweeping the forest floor. This has meant a reduction of soil organic matter and has interfered with the process of nutrient recycling, that is essential for maintaining the productivity of the forest. The floor sweeping also has resulted in accelerated soil erosion and laterization of the soils.

**Grazing**. The forest are extensively used for grazing of livestock. Grass is collected for stall feeding, but animals are usually herded within forest areas, resulting in damage to both vegetation and soils. There is no attempt to improve the productivity of natural grazing through pasture improvement programs.

## Socio-Economic Characteristics

**Population**. The population comprises a number of sub-groups, both general and scheduled castes as well as quite a large number of tribal groups. The main forest dependent groups in the study area are the Santhals and the Layaks.

- **Santhals.** A forest dwelling tribe, the Santhals form the single largest tribal group in the district, with a distinct dialect and culture, that is cooperative in character. Traditionally forest dependent, they have used the forest more sustainably than most other communities and have generally taken a lead in establishing community protection.

- **Layaks.** A forest dependent social group who are labelled as a general caste, being neither tribals nor scheduled caste. Layak communities residing near forests have become adapt at using them for commercial gains. Some are engaged in making charcoal from unauthorized harvest of wood from forest areas. In particular, they have recently started making "malas" (bead chains) out of twigs of number of forest species for a living.

Both groups collect a variety of Non Wood Forest Products (NWFP) for consumption and sale. They are small land owners and depend on larger land owners for work as agricultural laborers, as well as for credit in times of scarcity.

- **Ruidas and Rais**. These are Scheduled caste groups that are less forest dependent than the Santhals and Layaks. While the Rais work mainly as agricultural laborers, the Ruidas have a more diversified occupational pattern, with animal slaughter and agriculture cultivation as important main or subsidiary occupations.

- **Ghoshis and Sadgopis**. These groups traditionally specialized in raising livestock and have now taken up agriculture. Ghoshis tend to be larger landowners than the other groups, employing agricultural laborers to work the land, and in at least one of the villages they have also become involved with marketing of animal products, running a thriving dairy business in Durgapur. Although not numerically in the majority in many villages, by both caste and wealth they tend to be dominant in the communities where they are found.

- **Other Groups**. These include Telis, Lohars, Gorai, Patro, and Modok.

**Agriculture**. Rainfed paddy is the principal crop, though with the availability of irrigation from the Kangsabati project, other crops such as vegetables and betel leaves are also being cultivated. A number of families in the region migrate for labor to other agricultural districts. Animal husbandry forms an important part of the farming system in the villages near the forest areas. Livestock are kept mainly for draft power and for the production of farm yard manure. The recent increase in cultivation of betel leaves in the area has created a heavy demand for manure. Production of milk, either for home consumption or sale, seems to be the least important reason for rearing livestock. Wealthier households hire labor for herding their livestock, but most households herd their own animals.

## Dependence on the Forests

The forest has traditionally been the primary resource for many of the groups within the area, in particular for the Santhals and Layaks, who had traditional forest protection systems. This has facilitated the formation of FPCs. Careful use of forest resources which does not harm the forest is accepted, though felling trees for household construction or agricultural needs requires the approval of the FPC. Where viable alternatives to use of forest resources have been found, particularly sources of cash income, dependence on the forest has decreased. Elsewhere, dependence on forests remains, though communities tend to exploit unprotected areas.

Domestic fuel, fodder and grasses are collected throughout the year. Extraction of timber to meet construction or agricultural needs is an occasional event. Collection of minor forest produce, either for direct sale or for making products for sale, tends on the whole to be seasonal. In areas where JFM is practiced, there has been a major change in the fuel consumption, as green felled wood has been replaced by leaf litter, twigs and dry wood, purchases from pole traders and saw mills, or by other types of fuel such as coal, kerosene and bio-gas. This has resulted in a decline in headloading. However, felling of green wood for as fuelwood has not completely stopped in the area and headloaders still bring green-felled fuelwood for sale in small towns, particularly to large scale users, such as restaurants. Timber for agricultural implements and house construction is also derived from the forest. Plough blades and poles and threshing poles are generally replaced every year, yokes after about five years, and levellers after about eight or nine years. Carts are made from sal, or siris timber and bamboo, and last for ten to fifteen years. Poles used for house construction, mainly for beams and rafters, have an average life span of about 20 years. Sal, chikolda, jamun, akashmani and bamboo are all used. There is some charcoal production, but data on the scale of the industry were not available.

As many of the forest dependent groups are landless or own very small holdings, there is some encroachment on forest lands. This is fairly small scale and little land is reported to have been encroached on since the formation of the FPCs.

NWFP are perceived as one of the major reasons for community interests in JFM. A number of studies show:

- marketable forest products are of the most importance and there has been a rising trend for the collection of these in the past ten to twelve years;

- the importance of forests as a source of food and for meeting subsistence needs other than for fuel and fodder is declining;

- the total current gains from the forests for individual households, including fuel and fodder supplies and grazing, is generally in the range of Rs 1,000 to Rs 2,500 per year and less than about 20% of the total computed income;

- the volume of collection of NWFPs is at a minimum during the early monsoon months between June and August, when the requirement for labor for agriculture is at a maximum.

**Community Perceptions of Incentives for JFM**

There are two different groups involved in forest protection. The first consists of forest dependent people who initiated a form of forest protection prior to the existence, and the promises, of JFM. They became active in protecting their forests to ensure a continuing supply of forest produce, persuading neighboring village groups to join them. They have a very real concern to protect the forest in order to ensure continued collection of the many NWFP which are an essential part of their subsistence and livelihood. These groups have become part of the FPCs which later became established within the context of JFM. Revenue from a share of the harvest is a valued incentive, but not all members of the group appear to be aware of this right, or perhaps, believe that they will receive their share.

The second group consists mainly of general caste, landowning people who are not so dependent on the forest, for whom the share in the harvest is the primary incentive for participating in forest protection. It is from this group that the many "passive" FPC members come. Because of their position, they have usually become the dominant group within FPC taking the lead in decisions regarding protection arrangements, selection of support activities to be undertaken within the area, and benefit sharing arrangements. The role played by this group acts as a disincentive to the first group, who see themselves as undertaking most of the protection activity but, at best, receiving the same share in harvests as those who do much less; in reality they expect to receive far less than their share. A summary of perceived incentives and disincentives to JFM is presented in the following table.

## Villagers' Perceptions of Incentives and Disincentives Relating to JFM

| Protection-Related activity | Benefits | Disincentives | | Comments |
|---|---|---|---|---|
| | | Observations | Coping Mechanisms | |
| Providing watchmen | * Discouragement of waste of forest resources (forest-dependent groups' primary benefit)<br>* 25% share in harvests (general caste and land-owner groups' primary benefit) | * Waste of effort if FD doesn't take action against offenders, especially wealthy and powerful<br>* Some offenders use violence; risk of being injured and unable to work or go to collect supplies in forest | * Some FPCs impose fines, so FD action only needed for problems<br>* those with no alternative to headloading go into further forests (outside own/ FPC area)<br>* FD has paid for care following injury incurred in patrolling | * Increase in minor offenders caught causes heavy extra load on FD which results in failure to follow up on all;<br>* Formal assignment of authority to collect FPC rights to patrol, and free FD staff to attend to major offences.<br>* Possibility of taking out insurance for FPC patrols could be explored.<br>* Not all forest-dependent groups even aware of rights to share in harvest under JFM |
| Protection of plantations | * Access to plantations for leaf litter and forest sweepings for fuel<br>* Many aware of benefits accruing to FPC from cashew plantation harvest within district | * Discourages illicit use of plantations for fuelwood<br>* "passive", more powerful FPC members, who do little to protect the forest will at best take equal shares from benefits, probably more than the groups which do most protecting. | * Minimise use of wood; rely mainly on forest sweepings and supplement when necessary with coal, offcuts from approved logging<br>* Go further afield to collect wood outside own/FPC area<br><br>* Some FPCs keep a roster, so input of forest-dependent can be duly recognised | * Effect of extensive shift to dependence on forest sweepings for fuel is damaging nutrient layer, causing poor growth especially of ground vegetation also needed (NWFPs) and mushrooms, leading to erosion and poor tree growth in long run<br>* Few FPCs keep such a roster: leaders do not see any need. |

| Protection-Related activity | Benefits | Disincentives | | Comments |
| --- | --- | --- | --- | --- |
| | | Observations | Coping Mechanisms | |
| Employment in Plantations | * Possibility of earning wages on plantation operations of FD within FPC area | | | Little mention of this by villagers |
| Harvesting | * Right to share in benefits | * Delays in approving harvest of older plantations protected for years before JFM;<br>* FPCs want auction of standing trees; fear delays in moving stock from FD depots will diminish value | | * Working Plan 20 years old; doesn't allow for forest protection effects and actual harvesting needs. Revision necessary in the light of JFM effects |
| Protection of standing forest allocated to FPC | * Right to a share in cleaning and harvest operations | * Already been taking responsibility; allocated areas sometimes less than actual area the self-initiated (forest dependent) groups see themselves as responsible for | | * sub groups dependent on forest more interested in protecting for livelihood and subsistence needs; non-allocation to their FPC = loss of access to areas they have 'atak' for; other groups see lesser area as reduction in potential sharing. |

| Protection-Related activity | Benefits | Disincentives — Observations | Disincentives — Coping Mechanisms | Comments |
|---|---|---|---|---|
| **Support Activities** | | | | * Care needed to ensure that most forest dependent groups have access to facilities created through support activities |
| Water management (irrigation, weels, fish tanks) | * Increased productivity of agricultural resources reduces dependence on the forests | * General castes and powerful members of FPC are often favoured in choice of location | | |
| Seed for sericulture | * Good market for tussore cocoons | * With growth of sal trees, harvest of cocoons becomes more difficult | | * LAMPS was engaged in pro-duction of cloth, but subsidy came to an end so activity also stopped. * Involvement in processing would increase returns for cocoons harvested; special plantations would make collection easier. |
| Leaf plate machines | * Allows villagers to gain value addition on sal leaves more easily | * Leaves harder to reach with better growth in protected areas; * Market increasingly limited; competition in quality and price from Orissa | * Seek alternative sources of cash income (e.g. making malas, charcoal, in Layak communities) | * Low quality leaves and low quality plates limit the market, especially when many households compete for the market. |
| Community facilities | * Increases community assets | * In multi-village FPCs, dominant village/group benefits; others unlikely to be site of such | | * Care needed to ensure com-munity facilities not only built in village with most EC members |
| LAMPS monopoly on tendu and sal production | | * LAMPS prices seen to be held low ("robbing the poor"); fresh tendu leaves taken at lower rates; * Sal seeds paid for in kind (salt); * LAMPS relatively inactive for sal leaves/plates | * Sell to (unauthorised) other traders who take dried tendu leaves at higher rates * sell some of seeds to traders | |

60

## Annex 5

### Economic Valuation. Pricing

**Mixed Teak Forest System - Gujerat**

**Current Stumpage Prices.** These were derived from market prices for the main forest products, adjusted for the costs of extraction.

- **Timber** The most valuable species are teak and seshan; prices depend on quality classes. Most of the teak in the survey areas was either class 'B' or 'C'. Prices from both private traders and from the FD depot in Rajpiplia were collected. A weighted average of prices for non-teak timber was estimated (Tables 1 and 2).

- **Poles** Two sizes of pole were priced: small (girth 0.25m, height 3m) and large (girth .30m, height 4m) and volume-equivalent prices calculated. Pole prices were obtained from the few auctions which had taken place in Rajpiplia (Table 3).

- **Fuelwood** Between 20% and 30% of the biomass harvested is in the form of fuelwood. Fuelwood prices were calculated for teak and other species separately. There is a large volume of teak 'fuelwood',which ends up being used for purposes other than fuelwood. As a result the mean price of so-called teak fuelwood is about ten times the price of non-teak fuelwood. The price used for the study was the FD depot price. (Table 4).

A summary of economic and financial prices is given below.

| Species | Class | Market price (Rs/m^3) | Stumpage value (Rs/m^3) |
|---|---|---|---|
| Teak (private dealers) | A B C | 27,000 15,000 8,750 | 20,104 11,400 6,234 |
| Teak (FD depot) | B/C | 11,500 | 8,324 |
| Teak poles | Small Large | 4,020 4,230 | 2,699 2,844 |
| Teak 'fuelwood' (FD upset rate) | | 1,680 | 1,156 |
| Seshan | A B | 15,700 11,375 | 11,516 8,645 |
| Sirish Sadar etc Bad Other | | 9,400 7,500 5,100 3,600 | 6,728 5,284 3,846 2,320 |
| Mean (Non-teak) | | | 4,272 |
| Fuelwood (FD depot) | | 225 | 108 |

**Headloaded Fuelwood.** Headloaded fuelwood is required for home consumption and sale in nearby urban and semi-urban areas. Sale prices for headloads in different places show considerable variation ranging from Rs 7 to Rs 15 for one headload (15-20kg). This depends on both supply and demand. The labour content, valued at local wage rates, of one headload transported to a urban area exceeds its sale price. On this basis the opportunity cost of labour is less than Rs 10.00 per day (Table 5).

Fuelwood is also sold by the FD through a depot in Rajipiplia. The sale price is Rs 75.00 per quintel, about 75% of the reported market rate in Rajpiplia. However, even at this rate, supply exceeds demand. A timber merchant in Rajpiplia used to supply fuelwood as by-product at Rs 60 per quintel, but regards this as uneconomic; another supplies to a local Ashram at Rs 35.00 per quintel. The conclusion of this analysis is that fuelwood, itself, has little or no economic value in the field but provides a means of income to headloaders (valued at Rs 135/quintel).

**Bamboo**. The FD 'upset' rate for bamboo is Rs 8 per pole. There is also a concessional rate of Rs 1.25 per pole for (a) tribal vilages for home repairs (50 poles per year), (b) Kotwalis (800 poles per year) and, (c) Bansphiliol (20 poles per year, if less than 5 km from forest and 50 poles per year if between 5 and 15 km away). Extraction of bamboo is contracted out to a private sector paper mill (CPM) at agreed rates and for a fixed level of supply. Harvesting costs are Rs 45.00 per 100 poles. CPM are interested in 'lops and tops' of bamboo (Rs 70.00 per tonne 'green' weight and Rs 90 per tonne 'dry' weight). One pole is associated with 8.50 kg of 'lops and tops'. Without JFM, less bamboo ends up as pole quality. The value of Rs 2 per pole, therefore, has been adjusted to reflect a considerably higher proportion of lops and tops.

**Grass**. The variation in the demand and supply of grass relates to seasonal factors. In winter grass is cut and sold for storage at Rps 0.60 per bundle. In summer at peak demand, the same bundle, albeit drier, will be sold for Rs 0.80 per bundle. The weight of a bundle in winter varies from 0.75kg to 1 kg, say .80 kg, giving a sale price of Rs 75 per quintel. Grass harvesting data from an FLC shows that 166,000 kg of grass can be cut by 250 people working for 15 days - roughly 44 kg per person per day. This is equivalent to Rs 33 per day, which after deducting the labour cost of Rs 15 per day, gives an economic value of Rs 18 per 44 kg; or Rs 410 per tonne. This represents the value of grass in the field whether grazed or cut
(Table 6).

**Tendu**. Tendu is paid by the standard bag of 40 kg. After payment of collection and marketing costs, including a royalty of about 2% of sale price, the value of leaves on the bush/tree is about Rs 2.40 per kg. Assuming that leaves are collected over 50% of the forest area in the range, gives a mean collection rate of 14 kg/ha and a mean value per ha of Rs 33.60. The value of Rs 33.60 per ha is relatively low; however, even if it was assumed that collection was over a more limited area, the value per ha would not rise significantly - if tendu collection was restricted to 10% of the forest area, the value per ha would rise to Rs 167 per ha per year.

**Amla**.           The price of amla, a proxy for fruit in general, is assumed to be Rs 2.5 per kg.

A summary of the prices (Table MT.30) used to value products is given in the table below.

| Product | Unit | Market | 'Stumpage' |
|---|---|---|---|
| Teak    Wood | m$^3$ | 11,500.00 | 8,324.00 |
| Poles -small | pole | 4,020.00 | 2,699.00 |
| Poles -large | pole | 4,230.00 | 2,844.00 |
| Fuelwood | m$^3$ | 1,680.00 | 1,156.00 |
| Non-teak wood | m$^3$ | 6,100.00 | 4,272.00 |
| Fuelwood headloaded | qtl | 135.00 | 108.00 |
| Bamboo Good qlty | culm | 7.55 | 8.00 |
| Mixed qlty | culm | 1.55 | 2.00 |
| Lops/tops | kg | 0.11 | 0.11 |
| Grazing | ton | 460.00 | 410.00 |
| Tendu | Rs/ha | 33.60 | 33.60 |
| Fruit | kg | 2.50 | 2.50 |

**Price Trends**. The value of forestry products depends on **future**, rather than current, prices. Current prices have been taken as proxies for future prices; these probably best reflect FPC and FD perceptions of value and are therefore appropriate in an analysis of incentives.

**The Sal Coppice Forest System - West Bengal**

**Current Stumpage Prices**.

- **Sal poles**. The stumpage value of sal poles was estimated using market prices, obtained from dealers for different girth classes and using extraction and other standard costs obtained from a Forest Labour Cooperative (FLC). A weighted average price, based on the distribution of poles girths expected at different stages of the harvesting cycle, was calculated (Table 7).

- **Eucalyptus poles**. The stumpage value of eucalyptus poles was obtained using a similiar procedure (Table 8).

- **Fuelwood**. About 40% of the biomass harvested is in the form of fuelwood. Fuelwood prices were calculated for sal and other species separately (Tables 7 and 8).

- **NWFP.** Market prices of NWFPs were obtained from collectors, and where possible traders, in the field, along with quantities collected. Market prices were then adjusted to obtain equivalent 'stumpage' values (Table 9).

A summary of prices used to value products is given below.

|  | Product | Unit | Financial prices | Economic/ 'stumpage' prices |
|---|---|---|---|---|
| Sal | Fuelwood | Rs/m3 | 385 | 277 |
|  | Poles (7) | Rs/m3 | 1,388 | 1,211 |
|  | Poles (7/10) | Rs/m3 | 1,638 | 1,477 |
|  | Poles (10) | rs/m3 | 1,763 | 1,610 |
|  | Poles (16) | Rs/m3 | 2,305 | 2,204 |
|  | Poles (20) | Rs/m3 | 1,957 | 1,891 |
| Eucalyptus | Fuelwood | Rs/m3 | 375 | 239 |
|  | Biomass | Rs/m3 | 467 | 369 |
| NWFP | Mushroms | Rs/kg | 12 | 9.33 |
|  | Tendu | Rs/std bag | 200 | 130 |
|  | 'Other' | Rs/ha | 2,061 | 699 |
| Wage rates | FD | Rs/day | 40 | 28 |
|  | Agricultural | Rs/day | - | 15 |

**Planting Costs**

Planting costs for both natural forest and plantations were based on the current project as applied in the district (Tables 10 and 11).

**Price Trends.** The value of forestry products depends on **future**, rather than current, prices. Current prices have been taken as proxies for future prices; these probably best reflect FPC and FD perceptions of value and are therefore appropriate in an analysis of incentives.

**Table 1**  **Market Prices and Stumpage Values**

| Timber Class | Teak | | | | | Sishan | |
| | Depot | Private Dealers (Rajpiplia) | | | Ahmedabad | Private Dealers | |
| | B/C | A | B | C | A/B | A | B |
|---|---|---|---|---|---|---|---|
| Market (Sale) price | 11,500 | 27,000 | 15,000 | 8,750 | 22,750 | 15,700 | 11,375 |
| Extraction costs | | | | | | | |
| Logging | 10 | 10 | 0 | 10 | 10 | 10 | 0 |
| Transport | 321 | 321 | 0 | 321 | 321 | 321 | 0 |
| Fashioning | 132 | 132 | 0 | 132 | 132 | 132 | 0 |
| Debarking | 57 | 57 | 0 | 57 | 57 | 57 | 0 |
| FLC costs (5%) | 575 | 1,350 | 750 | 438 | 1,138 | 785 | 569 |
| Total costs | 1,095 | 1,870 | 750 | 957 | 1,657 | 1,305 | 569 |
| Gross margin | 10,406 | 25,131 | 14,250 | 7,793 | 21,093 | 14,396 | 10,806 |
| Profit (20%) | 2,081 | 5,026 | 2,850 | 1,559 | 4,219 | 2,879 | 2,161 |
| Stumpage value | 8,324 | 20,104 | 11,400 | 6,234 | 16,874 | 11,516 | 8,645 |

**Table 2**                    **Market Prices - Other Species**

|  | Species | | | | Teak |
|---|---|---|---|---|---|
|  | Sirish | Sadar | Bad | Other | |
| Sale price (m^3) | 9,400 | 7,500 | 5,100 | 3,600 | 11,500 |
| **Extraction costs** | | | | | |
| Logging/Marking | 10 | 10 | 0 | 10 | 10 |
| Transport | 321 | 321 | 0 | 321 | 321 |
| Fashioning | 132 | 132 | 0 | 132 | 132 |
| Debarking | 57 | 57 | 0 | 57 | 57 |
| FLC costs (5%) | 470 | 375 | 255 | 180 | 575 |
| Total costs | 990 | 895 | 255 | 700 | 1,095 |
| Gross margin | 8,411 | 6,606 | 4,845 | 2,901 | 10,406 |
| Profit (20%) | 1,682 | 1,321 | 969 | 580 | 2,081 |
| Stumpage value | 6,728 | 5,284 | 3,876 | 2,320 | 8,324 |
| Volume (m^3) | 31 | 491 | 93 | 275 | 973 |
| **Weighted average** | | | | | |
| Stumpage value | | 4,272 | | | |
| Market price | | 6,110 | | | |

**Table 3**     **Teak Pole Market Prices**

| Class | Small | Large |
|---|---|---|
| Sale price | 60.00 | 90.00 |
| Extraction costs | | |
|    Harvesting | 0.58 | 0.95 |
|    Transport | 3.06 | 4.41 |
|    FLC costs (10%) | 6.00 | 9.00 |
|    Total costs | 9.64 | 14.36 |
| Gross margin | 50.36 | 75.64 |
| Profit (20%) | 10.07 | 15.13 |
| Stumpage value | 40.29 | 60.51 |
| Poles per m^3 | 67 | 47 |
|    Gross | 4,020 | 4,230 |
|    Net | 2,699 | 2,844 |

**Table 4**          **Fuelwood Market Prices**

|  | Teak | | Non-Teak |
|---|---|---|---|
|  | MT | m^3 | m^3 |
| Sale price | 5,600.00 | 1,680.00 | 225.00 |
| Extraction costs | | | |
|     Harvesting (30% of total) | 30.00 | 9.00 | 9.00 |
|     Transport | 140.00 | 42.00 | 42.00 |
|     Stacking | 53.00 | 15.90 | 15.90 |
|     FLC costs (10%) | 560.00 | 168.00 | 22.50 |
|     Total costs | 783.00 | 234.90 | 89.40 |
| Gross margin | 4,817.00 | 1,445.10 | 135.60 |
| Profit (20%) | 963.40 | 289.02 | 27.12 |
| Stumpage value | 3,853.60 | 1,156.08 | 108.48 |
| Volume | | 60,000 | 20,200 |

69

**Table 5**  **Fuelwood (Headloading)**

|  | Locations | | | Timber | FD |
|  | Rajpiplia | Dediapadi | BedaCo | Merchant | Depot |
|---|---|---|---|---|---|
| Sale price (headload) | 15 | 10 | 8 | | |
| Sale price (qtl) | 94 | 63 | 50 | 60 | 75 |
| Distance<br>-village to market (km) | 40 | 12 | 6 | | |
| -village to forest (km) | 3 | 3 | 3 | | |
| Time required (hours) | | 12 | 8 | | |
| No of trips for 1 qtl | | 6.25 | 6.25 | | |
| Time required for 1 qtl | | 75 | 50 | | |
| Returns per hour | | 0.83 | 1.00 | | |

**Table 6 (a)**      **Grass and Fodder Market Prices**

|  |  | Rps/bundle |  | Kg/bundle | Rps/kg |
|---|---|---|---|---|---|
| Forest grass | Winter | 0.60 |  | 0.80 | 0.75 |
|  | Summer | 0.80 |  | 1.00 | 0.80 |
|  |  | Min | Max |  |  |
| Paddy straw |  | 0.70 | 1.00 | 1.00 | 0.70 |
| Jowar straw |  | 1.10 | 1.50 | 1.00 | 1.10 |

**Table 6 (b)**      **Seasonal Demand and Distribution of Grass/Fodder**

|  |  | Dec-Feb Winter Agric residue | Mar-June Summer Stall feeding | July-Nov Monsoon Forest grazing |
|---|---|---|---|---|
| Sale price | Rs/qtl | 75 | 100 |  |
| Collection rate (1) | kg/day | 44 |  |  |
| No of days (qtl) | Rps/day | 2.26 |  |  |
| Wage rate (day) | Rps/day | 15 |  |  |
| Wages | Rps/qtl | 34 |  |  |
| Field value | Rs/qtl | 41 | 55 | 41 |
| No of months | JFM | 3 | 4 | 5 |
|  | Non-JFM | 3 |  | 9 |
| Mean value | JFM | Rs/qtl | 46 |  |
|  | Non-JFM | Rs/qtl | 41 |  |

Note: (1) 166,000 kg harvested over 15 days by 250 people

**Table 7 (a)**     **Sal Pole Price**

|  |  | Fuelwood | Pole (Diameter Class) | | | |
|---|---|---|---|---|---|---|
|  |  |  | in | 4 | 5 | 6 | 8 |
|  |  |  | cm | 10 | 13 | 15 | 20 |
| Height | ft |  |  | 16.00 | 16.00 | 16.00 | 16.00 |
|  | m |  |  | 4.88 | 4.88 | 4.88 | 4.88 |
| Volume | m^3 |  |  | 0.04 | 0.06 | 0.09 | 0.16 |
| Poles | No/m^3 |  |  | 25 | 16 | 11 | 6 |
| Price (agent) | Rs/pole |  |  | 61 | 174 | 186 | 369 |
|  | Rs/m^3 | 385 |  | 1,542 | 2,815 | 2,090 | 2,332 |
| Costs | Harvesting (Rs 7/pole) | 70 |  | 177 | 113 | 79 | 44 |
|  | FDC cost (10% gross) | 39 |  | 154 | 282 | 209 | 233 |
|  | FDC commission (10%net) | 0 |  |  |  |  |  |
|  | Total | 109 |  | 331 | 395 | 288 | 277 |
| Stumpage value |  |  |  |  |  |  |  |
|  | Rs/m^3 | 277 |  | 1,211 | 2,421 | 1,802 | 2,055 |
|  | Rs/pole |  |  | 48 | 150 | 160 | 325 |
| Stumpage plus Labor |  |  |  |  |  |  |  |
|  | Rs/m^3 | 347 |  | 1,388 | 2,534 | 1,881 | 2,099 |

**Table 7(b)**     **Size Class Harvestable in Different Years**

| Age (Year) | Proportions (%) by Diameter Class | | | | |
|---|---|---|---|---|---|
|  | in | 4 | 5 | 6 | 8 |
|  | cm | 10 | 13 | 15 | 20 |
| 7 |  | 100 |  |  |  |
| 10 |  | 65 | 31 | 4 |  |
| 15 |  |  | 65 | 35 |  |
| 20 |  |  |  | 65 | 35 |

**Table 7(c)**     **Weighted Average Prices of Harvestable Poles**

| Age (Year) | Weighted Average Price (Rs) | |
|---|---|---|
|  | Stumpage | Stumpage plus Labor |
| 7 | 1,211 | 1,388 |
| 10 | 1,610 | 1,763 |
| 15 | 2,204 | 2,305 |
| 20 | 1,891 | 1,957 |
| 7/10 | 1,477 | 1,638 |

## Table 8         Eucalyptus Pole Prices

| | | | | | Pole Class | | | |
|---|---|---|---|---|---|---|---|---|
| Girth | cm | 27.5 | 32.5 | 37.5 | 42.5 | 47.5 | 52.5 | 57.5 |
| Diameter | cm | 8.75 | 10.34 | 11.93 | 13.52 | 15.11 | 16.70 | 18.30 |
| Height | m | 4.88 | 4.88 | 4.88 | 4.88 | 4.88 | 4.88 | 4.88 |
| Volume | m^3 | 0.03 | 0.04 | 0.05 | 0.07 | 0.09 | 0.11 | 0.13 |
| Poles | No/m^3 | 34 | 24 | 18 | 14 | 11 | 9 | 8 |
| Price (agent) | pole | 11 | 24 | 45 | 90 | 115 | 135 | 150 |
| | m^3 | 375 | 586 | 825 | 1,284 | 1,314 | 1,263 | 1,170 |
| Costs | Harvesting (Rs 7/pole) | 239 | 171 | 128 | 100 | 80 | 65 | 55 |
| | FLC (10%of gross) | 37 | 59 | 82 | 128 | 131 | 126 | 117 |
| | Total | 276 | 229 | 211 | 228 | 211 | 192 | 172 |
| Stumpage value | | | | | | | | |
| | Rs/m^3 | 99 | 356 | 614 | 1,056 | 1,103 | 1,071 | 998 |
| | Rs/pole | 3 | 15 | 34 | 74 | 97 | 115 | 128 |
| Market price | Rs/pole | 7 | 20 | 35 | 75 | 90 | 120 | 130 |
| | Rs/m^3 | 239 | 488 | 642 | 1,070 | 1,028 | 1,122 | 1,014 |

| Product | | Fuelwood | Pulp | Pole | | | | |
|---|---|---|---|---|---|---|---|---|
| At age 7 | Proportion (%) | 30 | 35 | 35 | | | | |

| Weighted average | Economic | Rs/m^3 | 369 |
|---|---|---|---|
| | Market | Rs/m^3 | 467 |

| Pulpwood | | | |
|---|---|---|---|
| Price at mill | MT | | 650 |

Note: Pole classes converted from inches

## Table 9 (a)   NTFP: Market and Economic Prices

|  |  | Sal Plates ('000 plates) | Sal Seeds (kg) | Mushrooms (kg) | Kendu (std bag) | Tussar (piece) |
|---|---|---|---|---|---|---|
| Sale price | Rps/unit | 27.50 | 1.50 (2) | 12.00 | 200.00 | 0.50 |
| Collection rate | Units/day | 500 (1) | 10 | 1.50 | 7 | 60 |
| Collection costs | Rps/unit | 25.00 | 1.00 | 2.67 | 70.00 | 0.17 |
| Economic value | Rps/unit | 2.50 | 0.50 | 9.33 | 130.00 | 0.33 |
|  | Rps/kg |  |  |  | 6.50 |  |

| Wage | Rps/day | 10 |
|---|---|---|

**Notes:** (1) Includes adult + child

(2) Exchanged for 1.2 kg salt at LAMPS

## Table 9 (b)   Summary of Financial and Economic Prices

|  | Unit | Market Price | Return to Land | Explicit Conversion Factor (CF) |
|---|---|---|---|---|
| Sal Plates | '000 plates | 27.50 | 2.50 | 0.09 |
| Sal Seeds | kg | 1.50 | 0.50 | 0.33 |
| Tussar | piece | 0.50 | 0.33 | 0.67 |
| Other | Rs |  |  | 0.36 |
| Mushrooms | kg | 12.00 | 9.33 | 0.78 |
| Kendu | stdbag | 200.00 | 130.00 | 0.65 |

Note:  Pole classes converted from inches

# Table 10 — Sal Forest Regeneration Unit Planting Costs in Financial Values

| | Unit | Rate | Year | | | | | | | | | | |
|---|---|---|---|---|---|---|---|---|---|---|---|---|---|
| | | | 0 | 1 | 2 | 3 | 4 | 5 | 6 | 7 | 8 | 9 | 10 |
| **Labor** | | | | | | | | | | | | | |
| Survey and demarcation | Day | 39.58 | 1 | | | 1 | | | | 1 | | | |
| Digging boundary trenchs | Day | 39.58 | 2 | | | 2 | | | | 2 | | | |
| Cutting sal shoots & cleaning | Day | 39.58 | 11 | | | | | | | | | | |
| Marking lead shoots | Day | 39.58 | | | | 3 | | | | 3 | | | |
| Cutting dead/diseased | Day | 39.58 | | | | 7.5 | | | | 8 | | | |
| Digging V trench | Day | 39.58 | 3.5 | | | | | | | 3.5 | | | |
| Watch and ward | Year | 3600 | | 0.03 | 0.03 | 0.03 | 0.03 | 0.03 | 0.03 | 0.03 | 0.03 | 0.03 | 0.03 |
| Sub-total | | 1,843 | 693 | 120 | 120 | 654 | 120 | 120 | 120 | 813 | 120 | 120 | 120 |
| (excluding harvesting costs) | | 1,138 | 257 | 120 | 120 | 357 | 120 | 120 | 120 | 496 | 120 | 120 | 120 |
| **Materials** | | | | | | | | | | | | | |
| Misc | LS | | 10 | | | 10 | | | | 10 | | | |
| Sub-total | | | 10 | 0 | 0 | 10 | 0 | 0 | 0 | 10 | 0 | 0 | 0 |
| **Total** | | | 703 | 120 | 120 | 664 | 120 | 120 | 120 | 823 | 120 | 120 | 120 |
| (excluding harvesting costs) | | | 267 | 120 | 120 | 367 | 120 | 120 | 120 | 506 | 120 | 120 | 120 |

Notes
Based on Treatment model R1; Forest regeneration of viable rootstock; PIV for World Bank funded Forestry Project
Wage rates: FD Rs 39.58 per day; Peak agricultural Rs 28.00 per day

**Table 11**  **Eucalyptus/Acacia Unit Plantation Costs Financial Terms**

| Inputs and costs | Unit | Rate | Year | | | | | | | |
|---|---|---|---|---|---|---|---|---|---|---|
| | | | 0 | 1 | 2 | 3 | 4 | 5 | 6 | 7 |
| Labor | | | | | | | | | | |
| Nursery operations | Day | 40 | 10 | 25 | | | | | | |
| Survey, cleaning, allignment | Day | 40 | 5 | | | | | | | |
| Digging v trench | Day | 40 | 10 | | | | | | | |
| Digging pits | Day | 40 | 15 | 30 | | | | | | |
| Filling pits | Day | 40 | | 10 | | | | | | |
| Planting with potted seedlings | Day | 40 | | 15 | | | | | | |
| Weeding | Day | 40 | | 25 | | | | | | |
| Watch and ward | Year | 3600 | | 0.03 | 0.03 | 0.03 | 0.03 | 0.03 | | |
| Sub-total | | | 1600 | 4320 | 120 | 120 | 120 | 120 | | |
| Materials | | | | | | | | | | |
| Shed and fencing | LS | | 40 | 20 | | | | | | |
| FYM | LS | | 80 | | | | | | | |
| Seeds | LS | | 20 | 10 | | | | | | |
| Watering cans etc | LS | | 10 | | | | | | | |
| Misc | LS | | 10 | | | | | | | |
| Sub-total | | | 160 | 30 | 0 | 0 | 0 | 0 | | |
| Total | | 5,330 | 1760 | 4350 | 120 | 120 | 120 | 120 | | |
| Outputs and benefits | | | | | | | | | | |
| Poles/fuelwood | m^3/ha | 466.97 | | | | | | | | 28 |
| Tendu | std bag/ha | 200.00 | 0.26 | 0.26 | 0.26 | 0.26 | 0.26 | 0.26 | 0.26 | 0.26 |
| Total | | 5,543 | 53 | 53 | 53 | 53 | 53 | 53 | 53 | 13,128 |

**Table MT.1 Forest Area**

| | Area (ha) | | |
|---|---|---|---|
| | FPC | Forest Dept | Total |
| Natural forest (NF) | 90.0 | 95.0 | 185.0 |
| Plantation forest (PL) | 107.5 | | 107.5 |
| Grazing land | 7.5 | | 7.5 |
| Total | 205.0 | 95.0 | 300.0 |

**Table MT.2    Sample Plots**

| Type | Degradation Status | No | Species | Sample Area (ha) |
|---|---|---|---|---|
| Natural Forest | Moderate | NF.1 | Mixed | 30.0 |
| Natural Forest | Moderate | NF.2 | Mixed | 20.0 |
| Natural Forest | Partial | NF.3 | Mixed | 20.0 |
| Natural Forest | Highly | NF.4 | Mixed | 20.0 |
| Sub-total | | | | 90.0 |
| Grazing land | Highly | | | 7.5 |
| Sub-total | | | | 7.5 |
| Plantation | | PL.1 | Teak | 2.5 |
| Plantation | | PL.2 | Sevan | 5.0 |
| Plantation | | PL.3 | Subabul | 15.0 |
| Plantation | | PL.3 | SWC | 10.0 |
| Plantation | | PL.3 | MFP | 15.0 |
| Plantation | | PL.4 | Fodder | 10.0 |
| Plantation | | PL.5 | Waste | 25.0 |
| Plantation | | | Mixed | 25.0 |
| Sub-total | | | | 107.5 |
| Total | | | | 205.0 |

**Table MT.3    Silvicultural Parameters for Natural Forest**

| Sample Area | Area (ha) | Protection | Soil | Species | | Stems (No/ha) | Basal Area (m^2/ha) | Height (m) | Girth (cm) | Form Factor | Growing stock (m^3/ha) | Age (years) | MAI (m^3/ha/yr) |
|---|---|---|---|---|---|---|---|---|---|---|---|---|---|
| NF.1 | 30 | Good | Black cotton | Teak | | 10 | | 9.0 | 25-30 | 0.382 | 2.95 | 25 | 0.12 |
| | | | | Non-teak | | 200 | | | | | 29.57 | | 1.18 |
| | | | | | Sadar | 50 | | 7.5 | 20-25 | 0.400 | 12.86 | 25 | 0.51 |
| | | | | | Kakad | 50 | | 6.0 | 20-25 | 0.250 | 6.43 | 25 | 0.26 |
| | | | | | Kudi | 100 | | 6.0 | 15-20 | 0.200 | 10.29 | 25 | 0.41 |
| | | | | Total | | 210 | 18 | | | | 32.52 | | 1.30 |
| | | | | Bamboo | | 100 | | | 16 | | 1,600 | | |
| NF.2 | 20 | Moderate | Black cotton | Teak | | 10 | | 9.0 | 25-30 | 0.382 | 1.96 | 25 | 0.08 |
| | | | | Non-teak | | 200 | | | | | 19.71 | | 0.79 |
| | | | | | Sadar | 50 | | 7.5 | 20-25 | 0.400 | 8.57 | 25 | 0.34 |
| | | | | | Kakad | 50 | | 6.0 | 20-25 | 0.250 | 4.29 | 25 | 0.17 |
| | | | | | Kudi | 100 | | 6.0 | 15-20 | 0.200 | 6.86 | 25 | 0.27 |
| | | | | Total | | 210 | 12 | | | | 21.68 | | 0.87 |
| | | | | Bamboo | | 100 | | | 16 | | 1,600 | | |
| NF.3 | 20 | Moderate | Sandy loam | Teak | | 900 | | 9.0 | 20-30 | 0.382 | 16.14 | 15 | 1.08 |
| | | | | Non-teak | | 1400 | | | | | 11.43 | | 0.76 |
| | | | | | Sadar | 100 | | 7.5 | 15-20 | 0.400 | 1.57 | 15 | 0.10 |
| | | | | | Kakad | 1100 | | 6.0 | 15-20 | 0.250 | 8.61 | 15 | 0.57 |
| | | | | | Others | 200 | | 6.0 | 12-15 | 0.200 | 1.25 | 15 | 0.08 |
| | | | | Total | | 2300 | 12 | | | | 27.57 | | 1.84 |
| | | | | Bamboo | | 100 | | | 16 | | 6,400 | | |
| NF.4 | 20 | Moderate | Sandy loam | Teak | | 400 | | 9.0 | 15-20 | 0.382 | 8.60 | 15 | 0.57 |
| | | | | Non-teak | | 1200 | | | | | 10.31 | | 0.69 |
| | | | | | Sadar | 100 | | 7.5 | 10-15 | 0.400 | 1.88 | 15 | 0.13 |
| | | | | | Kakad | 100 | | 6.0 | 10-15 | 0.250 | 0.94 | 15 | 0.06 |
| | | | | | Kudi | 1000 | | 6.0 | 10-12 | 0.200 | 7.50 | 15 | 0.50 |
| | | | | Total | | 1600 | 10 | | | | 18.91 | | 1.26 |
| | | | | Bamboo | | 400 | | | 4 | | 1,600 | | |

Notes:    Growing Stock (Species) = Proportion of stems x height x Basal Area x Form Factor

MAI = Growing Stock/Age

Bamboo data in culms

**Table MT.4    Forest Productivity Model (Type NF-1)**

| Management | | Output | Unit | Unit Price (Rs) | Offtake Rates (% CAI) | Year 0 | 1 | 2 | 6 | 10 | 14 | 18 | 22 | 25 | 30 | 35 |
|---|---|---|---|---|---|---|---|---|---|---|---|---|---|---|---|---|
| **WITH JFM** | | | | | | | | | | | | | | | | |
| | | GS | M^3/ha | | | 32.52 | 35.10 | 37.20 | 44.16 | 49.25 | 52.98 | 55.70 | 57.69 | 58.83 | 0.00 | 0.00 |
| | | CAI | M^3/ha | | | 2.58 | 2.36 | 2.19 | 1.60 | 1.17 | 0.86 | 0.63 | 0.46 | 0.36 | 0.00 | 0.00 |
| | | CC | (%) | | | 30 | 34 | 37 | 48 | 58 | 65 | 71 | 75 | 78 | 0 | 0 |
| | | Dead/dry | M^3/ha | 135.00 | 10 % | | 0.26 | 0.24 | 0.17 | 0.13 | 0.09 | 0.07 | 0.05 | 0.04 | 0.00 | 0.00 |
| | | Green cut | M^3/ha | 135.00 | 0 % | | 0.00 | 0.00 | 0.00 | 0.00 | 0.00 | 0.00 | 0.00 | 0.00 | 0.00 | 0.00 |
| | | Teak | M^3/ha | 8,324.00 | | | | | | | | | | 2.24 | | |
| | | Fuelwood | M^3/ha | 1,152.00 | | | | | | | | | | 0.56 | | |
| | | Non teak | M^3/ha | 459.00 | | | | | | | | | | 39.22 | | |
| | | Fuelwood | M^3/ha | 135.00 | | | | | | | | | | 16.81 | | |
| | | Bamboo | Poles | 8.00 | | 100 | | 800 | 800 | 800 | 800 | 800 | 800 | | 800 | |
| | | Lops/tops | kg/pole | 0.11 | | 8.50 | | 6,800 | 6,800 | 6,800 | 6,800 | 6,800 | 6,800 | | 6,800 | |
| | | Grazing | MT/ha | 460.00 | | | 0.73 | 0.66 | 0.43 | 0.24 | 0.10 | -0.02 | -0.10 | -0.15 | 1.40 | 1.40 |
| | | Tendu | LS/ha | 1 | | | 60.00 | 60.00 | 40.00 | 40.00 | 40.00 | 40.00 | 40.00 | 40.00 | 120.00 | 120.00 |
| | | NTFP | | 1 | | | 150.00 | 150.00 | 250.00 | 350.00 | 450.00 | 450.00 | 450.00 | 450.00 | 0.00 | 0.00 |
| **Summary of benefits with JFM** | | Net Worth (Rs) | | Discount Rate (%) | | | | | | | | | | | | |
| FD | Revenue | 8,863 | | 12 | | | 0 | 3,581 | 3,581 | 3,581 | 3,581 | 3,581 | 3,581 | 19,786 | 3,581 | 0 |
| FPC | Revenue | 8,863 | | 12 | | | 0 | 3,581 | 3,581 | 3,581 | 3,581 | 3,581 | 3,581 | 19,786 | 3,581 | 0 |
| | Headloaders | 176 | | 12 | | | 35 | 32 | 23 | 17 | 12 | 9 | 7 | 5 | 0 | 0 |
| | Graziers | 1,609 | | 12 | | | 334 | 304 | 198 | 112 | 44 | -8 | -47 | -70 | 642 | 642 |
| | Collectors | 2,743 | | 12 | | | 210 | 210 | 290 | 390 | 490 | 490 | 490 | 490 | 120 | 120 |
| | Total | 13,391 | | 12 | | | 578 | 4,127 | 4,092 | 4,100 | 4,127 | 4,072 | 4,030 | 20,211 | 4,343 | 762 |
| Economic | | 22,255 | | 12 | | | 578 | 7,708 | 7,673 | 7,680 | 7,708 | 7,653 | 7,611 | 39,997 | 7,924 | 762 |
| **WITHOUT JFM** | | | | | | | | | | | | | | | | |
| | | GS | M^3/ha | | | 32.52 | 35.10 | 36.69 | 42.26 | 46.66 | 50.13 | 52.87 | 55.03 | 56.35 | 58.08 | 59.38 |
| | | CAI | M^3/ha | | | 2.58 | 2.36 | 2.23 | 1.76 | 1.39 | 1.10 | 0.86 | 0.68 | 0.57 | 0.42 | 0.32 |
| | | CC | (%) | | | 30 | 34 | 36 | 45 | 53 | 60 | 65 | 69 | 72 | 76 | 79 |
| | | Dead/dry | M^3/ha | 135.00 | 10 % | | 0.26 | 0.24 | 0.19 | 0.15 | 0.12 | 0.09 | 0.07 | 0.06 | 0.05 | 0.03 |
| | | Green cut | M^3/ha | 135.00 | 20 % | | 0.52 | 0.47 | 0.37 | 0.29 | 0.23 | 0.18 | 0.14 | 0.12 | 0.09 | 0.07 |
| | | Teak | M^3/ha | 8,324.00 | | | | | | | | | | | | 2.26 |
| | | Fuelwood | M^3/ha | 1,152.00 | | | | | | | | | | | | 0.57 |
| | | Non teak | M^3/ha | 459.00 | | | | | | | | | | | | 39.58 |
| | | Fuelwood | M^3/ha | 135.00 | | | | | | | | | | | | 16.96 |
| | | Bamboo | Poles | 2.00 | | 100 | 200 | 200 | 200 | 200 | 200 | 200 | 200 | 200 | 200 | 200 |
| | | Lops/tops | kg/pole | 0.11 | | 4.25 | 850 | 850 | 850 | 850 | 850 | 850 | 850 | 850 | 850 | 850 |
| | | Grazing | MT/ha | 410.00 | | | 0.73 | 0.68 | 0.50 | 0.34 | 0.21 | 0.10 | 0.01 | -0.04 | -0.12 | -0.18 |
| | | Tendu | LS/ha | 1 | | | 60.00 | 60.00 | 40.00 | 40.00 | 40.00 | 40.00 | 40.00 | 40.00 | 40.00 | 40.00 |
| | | NTFP | | 1 | | | 150.00 | 150.00 | 250.00 | 350.00 | 350.00 | 450.00 | 450.00 | 450.00 | 450.00 | 450.00 |
| **Summary of benefits without JFM** | | Net Worth (Rs) | | Discount Rate (%) | | | | | | | | | | | | |
| FD | | 4,791 | | 12 | | | 494 | 494 | 494 | 494 | 494 | 494 | 494 | 494 | 494 | 40,432 |
| FPC | Revenue | 0 | | 12 | | | 0 | 0 | 0 | 0 | 0 | 0 | 0 | 0 | 0 | 0 |
| | Headloaders | 574 | | 12 | | | 104 | 96 | 76 | 60 | 47 | 37 | 29 | 25 | 18 | 14 |
| | Graziers | 1,436 | | 12 | | | 297 | 278 | 203 | 139 | 86 | 41 | 4 | -18 | -49 | -72 |
| | Collectors | 2,654 | | 12 | | | 210 | 210 | 290 | 390 | 390 | 490 | 490 | 490 | 490 | 490 |
| | Total | 4,664 | | 12 | | | 612 | 583 | 569 | 589 | 523 | 568 | 524 | 496 | 459 | 431 |
| Economic | | 9,455 | | 12 | | | 1,105 | 1,077 | 1,063 | 1,083 | 1,016 | 1,062 | 1,017 | 990 | 953 | 40,863 |
| **INCREMENTAL BENEFITS DUE TO JFM** | | Net Worth (Rs) | | Discount Rate (%) | | | | | | | | | | | | |
| FD | | 4,072 | | 12 | | | -494 | 3,087 | 3,087 | 3,087 | 3,087 | 3,087 | 3,087 | 19,292 | 3,087 | -40,432 |
| FPC | Revenue | 8,863 | | 12 | | | 0 | 3,581 | 3,581 | 3,581 | 3,581 | 3,581 | 3,581 | 19,786 | 3,581 | 0 |
| | Headloaders | -398 | | 12 | | | -70 | -64 | -52 | -43 | -35 | -28 | -23 | -19 | -18 | -14 |
| | Graziers | 174 | | 12 | | | 36 | 27 | -6 | -28 | -42 | -49 | -52 | -52 | 691 | 715 |
| | Collectors | 89 | | 12 | | | 0 | 0 | 0 | 0 | 100 | 0 | 0 | 0 | -370 | -370 |
| | Total | 8,728 | | 12 | | | -33 | 3,544 | 3,523 | 3,510 | 3,605 | 3,504 | 3,506 | 19,715 | 3,884 | 331 |
| Economic | | 12,800 | | 12 | | | -527 | 6,631 | 6,610 | 6,598 | 6,692 | 6,591 | 6,594 | 39,007 | 6,971 | -40,101 |

**Table MT.5    Forest productivity model (Natural Forest Type NF-2)**

| Management | Output | Unit | Unit Price (Rs) | Offtake rates (% CAI) | 0 | 1 | 2 | 6 | 10 | 14 | 18 | 22 | 25 | 30 | 35 |
|---|---|---|---|---|---|---|---|---|---|---|---|---|---|---|---|
| **WITH JFM** | | | | | | | | | | | | | | | |
| | GS | M^3/ha | | | 21.68 | 25.17 | 28.02 | 37.45 | 44.34 | 49.38 | 53.07 | 55.77 | 57.31 | 0.00 | 0.00 |
| | CAI | M^3/ha | | | 3.49 | 3.20 | 2.96 | 2.16 | 1.58 | 1.16 | 0.85 | 0.62 | 0.49 | 0.00 | 0.00 |
| | CC | (%) | | | 17 | 20 | 24 | 37 | 49 | 58 | 65 | 71 | 74 | 0 | 0 |
| | Dead/dry | M^3/ha | 135.00 | 10 % | | 0.35 | 0.32 | 0.23 | 0.17 | 0.13 | 0.09 | 0.07 | 0.05 | 0.00 | 0.00 |
| | Green cut | M^3/ha | 135.00 | 0 % | | 0.00 | 0.00 | 0.00 | 0.00 | 0.00 | 0.00 | 0.00 | 0.00 | 0.00 | 0.00 |
| | Teak | M^3/ha | 8,324.00 | | | | | | | | | | 2.18 | | |
| | Fuelwood | M^3/ha | 1,152.00 | | | | | | | | | | 0.55 | | |
| | Non teak | M^3/ha | 459.00 | | | | | | | | | | 38.21 | | |
| | Fuelwood | M^3/ha | 135.00 | | | | | | | | | | 16.37 | | |
| | Bamboo | Poles | 8.00 | | 100 | | 800 | 800 | 800 | 800 | 800 | 800 | | 800 | |
| | Lops/tops | kg/pole | 0.11 | | 8.50 | | 6,800 | 6,800 | 6,800 | 6,800 | 6,800 | 6,800 | | 6,800 | |
| | Grazing | MT/ha | 460.00 | | | 0.99 | 0.92 | 0.65 | 0.42 | 0.24 | 0.09 | -0.02 | -0.09 | 1.40 | 1.40 |
| | Tendu | LS/ha | 1 | | | 80.00 | 80.00 | 60.00 | 40.00 | 40.00 | 40.00 | 40.00 | 40.00 | 120.00 | 120.00 |
| | NTFP | | 1 | | | 50.00 | 50.00 | 150.00 | 250.00 | 350.00 | 450.00 | 450.00 | 450.00 | 0.00 | 0.00 |

**Summary of benefits with JFM**

| | | Net Worth (Rs) | Discount Rate (%) | | 0 | 1 | 2 | 6 | 10 | 14 | 18 | 22 | 25 | 30 | 35 |
|---|---|---|---|---|---|---|---|---|---|---|---|---|---|---|---|
| FD | | 8,833 | 12 | | | 0 | 3,581 | 3,581 | 3,581 | 3,581 | 3,581 | 3,581 | 19,274 | 3,581 | 0 |
| FPC | Revenue | 8,833 | 12 | | | 0 | 3,581 | 3,581 | 3,581 | 3,581 | 3,581 | 3,581 | 19,274 | 3,581 | 0 |
| | Headloaders | 238 | 12 | | | 47 | 43 | 32 | 23 | 17 | 12 | 9 | 7 | 0 | 0 |
| | Graziers | 2,350 | 12 | | | 456 | 423 | 301 | 195 | 109 | 42 | -9 | -40 | 642 | 642 |
| | Collectors | 2,134 | 12 | | | 130 | 130 | 210 | 290 | 390 | 490 | 490 | 490 | 120 | 120 |
| | Total | 13,555 | 12 | | | 633 | 4,177 | 4,123 | 4,089 | 4,097 | 4,125 | 4,071 | 19,732 | 4,343 | 762 |
| Economic | | 22,389 | 12 | | | 633 | 7,758 | 7,704 | 7,670 | 7,678 | 7,706 | 7,651 | 39,006 | 7,924 | 762 |

| Management | Output | Unit | Unit Price (Rs) | Offtake rates (% CAI) | 0 | 1 | 2 | 6 | 10 | 14 | 18 | 22 | 25 | 30 | 35 |
|---|---|---|---|---|---|---|---|---|---|---|---|---|---|---|---|
| **WITHOUT JFM** | | | | | | | | | | | | | | | |
| | GS | M^3/ha | | | 21.68 | 25.17 | 26.62 | 32.17 | 36.88 | 40.87 | 44.25 | 47.12 | 48.98 | 51.61 | 53.76 |
| | CAI | M^3/ha | | | 3.49 | 3.20 | 3.08 | 2.61 | 2.21 | 1.88 | 1.59 | 1.35 | 1.19 | 0.97 | 0.79 |
| | CC | (%) | | | 17 | 20 | 22 | 30 | 36 | 43 | 49 | 54 | 57 | 62 | 67 |
| | Dead/dry | M^3/ha | 135.00 | 10 % | | 0.35 | 0.32 | 0.27 | 0.23 | 0.20 | 0.17 | 0.14 | 0.12 | 0.10 | 0.08 |
| | Green cut | M^3/ha | 135.00 | 40 % | | 1.40 | 1.28 | 1.09 | 0.92 | 0.78 | 0.66 | 0.56 | 0.50 | 0.40 | 0.33 |
| | Teak | M^3/ha | 8,324.00 | | | | | | | | | | | | 2.05 |
| | Fuelwood | M^3/ha | 1,152.00 | | | | | | | | | | | | 0.51 |
| | Non teak | M^3/ha | 459.00 | | | | | | | | | | | | 35.84 |
| | Fuelwood | M^3/ha | 135.00 | | | | | | | | | | | | 15.36 |
| | Bamboo | Poles | 2.00 | | 100 | 200 | 200 | 200 | 200 | 200 | 200 | 200 | 200 | 200 | 200 |
| | Lops/tops | kg/pole | 0.11 | | 4.25 | 850 | 850 | 850 | 850 | 850 | 850 | 850 | 850 | 850 | 850 |
| | Grazing | MT/ha | 410.00 | | | 0.99 | 0.96 | 0.81 | 0.67 | 0.54 | 0.43 | 0.32 | 0.25 | 0.15 | 0.06 |
| | Tendu | LS/ha | 1 | | | 80.00 | 80.00 | 80.00 | 60.00 | 40.00 | 40.00 | 40.00 | 40.00 | 40.00 | 40.00 |
| | NTFP | | 1 | | | 50.00 | 50.00 | 50.00 | 150.00 | 250.00 | 250.00 | 350.00 | 350.00 | 450.00 | 450.00 |

**Summary of benefits without JFM**

| | | Net Worth (Rs) | Discount Rate (%) | | 0 | 1 | 2 | 6 | 10 | 14 | 18 | 22 | 25 | 30 | 35 |
|---|---|---|---|---|---|---|---|---|---|---|---|---|---|---|---|
| FD | | 4,719 | 12 | | | 494 | 494 | 494 | 494 | 494 | 494 | 494 | 494 | 494 | 36,653 |
| FPC | Revenue | 0 | 12 | | | 0 | 0 | 0 | 0 | 0 | 0 | 0 | 0 | 0 | 0 |
| | Headloaders | 1,409 | 12 | | | 236 | 216 | 184 | 156 | 132 | 112 | 95 | 84 | 68 | 56 |
| | Graziers | 2,459 | 12 | | | 406 | 392 | 332 | 275 | 223 | 175 | 133 | 104 | 62 | 26 |
| | Collectors | 1,662 | 12 | | | 130 | 130 | 130 | 210 | 290 | 290 | 390 | 390 | 490 | 490 |
| | Total | 5,530 | 12 | | | 772 | 738 | 645 | 641 | 645 | 577 | 617 | 578 | 620 | 572 |
| Economic | | 10,250 | 12 | | | 1,265 | 1,231 | 1,139 | 1,134 | 1,138 | 1,071 | 1,111 | 1,071 | 1,113 | 37,225 |

**INCREMENTAL BENEFITS DUE TO JFM**

| | | Net Worth (Rs) | Discount Rate (%) | | 0 | 1 | 2 | 6 | 10 | 14 | 18 | 22 | 25 | 30 | 35 |
|---|---|---|---|---|---|---|---|---|---|---|---|---|---|---|---|
| FD | | 4,114 | 12 | | | -494 | 3,087 | 3,087 | 3,087 | 3,087 | 3,087 | 3,087 | 18,781 | 3,087 | -36,653 |
| FPC | Revenue | 8,833 | 12 | | | 0 | 3,581 | 3,581 | 3,581 | 3,581 | 3,581 | 3,581 | 19,274 | 3,581 | 0 |
| | Headloaders | -1,171 | 12 | | | -189 | -173 | -152 | -133 | -115 | -100 | -86 | -77 | -68 | -56 |
| | Graziers | -109 | 12 | | | 50 | 32 | -31 | -80 | -113 | -133 | -142 | -143 | 581 | 616 |
| | Collectors | 472 | 12 | | | 0 | 0 | 80 | 80 | 100 | 200 | 100 | 100 | -370 | -370 |
| | Total | 8,025 | 12 | | | -139 | 3,440 | 3,478 | 3,448 | 3,452 | 3,548 | 3,453 | 19,154 | 3,723 | 191 |
| Economic | | 12,139 | 12 | | | -633 | 6,527 | 6,565 | 6,535 | 6,540 | 6,636 | 6,540 | 37,935 | 6,810 | -36,463 |

**Table MT.6    Forest Productivity Model (Type NF-3)**

| Management | Output | Unit | Unit Price (Rs) | Offtake rates (%CAI) | Year 0 | 1 | 2 | 6 | 10 | 14 | 18 | 22 | 25 | 30 | 35 |
|---|---|---|---|---|---|---|---|---|---|---|---|---|---|---|---|
| **WITH JFM** | | | | | | | | | | | | | | | |
| | GS | M^3/ha | | | 27.57 | 30.57 | 33.01 | 41.10 | 47.01 | 51.34 | 54.50 | 56.82 | 58.14 | 59.75 | 60.84 |
| | CAI | M^3/ha | | | 3.00 | 2.74 | 2.54 | 1.86 | 1.36 | 0.99 | 0.73 | 0.53 | 0.42 | 0.28 | 0.19 |
| | CC | (%) | | | 23 | 27 | 31 | 43 | 54 | 62 | 68 | 73 | 76 | 80 | 82 |
| | Dead/dry | M^3/ha | 135.00 | 10 % | | 0.30 | 0.27 | 0.20 | 0.15 | 0.11 | 0.08 | 0.06 | 0.05 | 0.03 | 0.02 |
| | Green cut | M^3/ha | 135.00 | 0 % | | 0.00 | 0.00 | 0.00 | 0.00 | 0.00 | 0.00 | 0.00 | 0.00 | 0.00 | 0.00 |
| | Teak | M^3/ha | 8,324.00 | | | | | | | | | | | | 19.05 |
| | Fuelwood | M^3/ha | 1,152.00 | | | | | | | | | | | | 4.76 |
| | Non teak | M^3/ha | 459.00 | | | | | | | | | | | | 25.92 |
| | Fuelwood | M^3/ha | 135.00 | | | | | | | | | | | | 11.11 |
| | Bamboo | Poles | 8.00 | | 400 | | 3,200 | 3,200 | 3,200 | 3,200 | 3,200 | 3,200 | | 3,200 | |
| | Lops/tops | kg/pole | 0.11 | | 8.50 | | 27,200 | 27,200 | 27,200 | 27,200 | 27,200 | 27,200 | | 27,200 | |
| | Grazing | MT/ha | 460.00 | | | 0.85 | 0.79 | 0.54 | 0.33 | 0.16 | 0.03 | -0.06 | -0.12 | -0.19 | -0.24 |
| | Tendu | LS/ha | 1 | | | 80.00 | 60.00 | 40.00 | 40.00 | 40.00 | 40.00 | 40.00 | 40.00 | 40.00 | 40.00 |
| | NTFP | | 1 | | | 50.00 | 150.00 | 250.00 | 350.00 | 450.00 | 450.00 | 450.00 | 450.00 | 450.00 | 400.00 |

**Summary of benefits with JFM**

| | | Net Worth (Rs) | Discount Rate (%) | | 0 | 1 | 2 | 6 | 10 | 14 | 18 | 22 | 25 | 30 | 35 |
|---|---|---|---|---|---|---|---|---|---|---|---|---|---|---|---|
| FD | | 32,478 | 12 | | | 0 | 14,323 | 14,323 | 14,323 | 14,323 | 14,323 | 14,323 | 0 | 14,323 | 88,712 |
| FPC | Revenue | 32,478 | 12 | | | 0 | 14,323 | 14,323 | 14,323 | 14,323 | 14,323 | 14,323 | 0 | 14,323 | 88,712 |
| | Headloaders | 205 | 12 | | | 40 | 37 | 27 | 20 | 15 | 11 | 8 | 6 | 4 | 3 |
| | Graziers | 1,719 | 12 | | | 393 | 361 | 246 | 151 | 74 | 15 | -30 | -56 | -89 | -112 |
| | Collectors | 2,569 | 12 | | | 130 | 210 | 290 | 390 | 490 | 490 | 490 | 490 | 490 | 440 |
| | Total | 36,971 | 12 | | | 563 | 14,932 | 14,887 | 14,884 | 14,902 | 14,839 | 14,791 | 440 | 14,728 | 89,043 |
| Economic | | 69,449 | 12 | | | 563 | 29,255 | 29,210 | 29,207 | 29,225 | 29,162 | 29,114 | 440 | 29,052 | 177,755 |

| Management | Output | Unit | Unit Price (Rs) | Offtake rates (%CAI) | Year 0 | 1 | 2 | 6 | 10 | 14 | 18 | 22 | 25 | 30 | 35 |
|---|---|---|---|---|---|---|---|---|---|---|---|---|---|---|---|
| **WITHOUT JFM** | | | | | | | | | | | | | | | |
| | GS | M^3/ha | | | 27.57 | 30.57 | 30.61 | 31.63 | 32.61 | 33.55 | 34.47 | 35.36 | 36.00 | 37.05 | 38.05 |
| | CAI | M^3/ha | | | 3.00 | 2.74 | 2.74 | 2.65 | 2.57 | 2.49 | 2.42 | 2.34 | 2.29 | 2.20 | 2.11 |
| | CC | (%) | | | 23 | 27 | 27 | 29 | 30 | 31 | 33 | 34 | 35 | 37 | 38 |
| | Dead/dry | M^3/ha | 135.00 | 10 % | | 0.30 | 0.27 | 0.27 | 0.26 | 0.25 | 0.24 | 0.24 | 0.23 | 0.22 | 0.21 |
| | Green cut | M^3/ha | 135.00 | 80 % | | 2.40 | 2.20 | 2.14 | 2.07 | 2.01 | 1.95 | 1.89 | 1.84 | 1.77 | 1.70 |
| | Teak | M^3/ha | 8,324.00 | | | | | | | | | | | | 11.91 |
| | Fuelwood | M^3/ha | 1,152.00 | | | | | | | | | | | | 2.98 |
| | Non teak | M^3/ha | 459.00 | | | | | | | | | | | | 16.21 |
| | Fuelwood | M^3/ha | 135.00 | | | | | | | | | | | | 6.95 |
| | Bamboo | Poles | 2.00 | | 400 | 800 | 800 | 800 | 800 | 800 | 800 | 800 | 800 | 800 | 800 |
| | Lops/tops | kg/pole | 0.11 | | 4.25 | 3,400 | 3,400 | 3,400 | 3,400 | 3,400 | 3,400 | 3,400 | 3,400 | 3,400 | 3,400 |
| | Grazing | MT/ha | 410.00 | | | 0.85 | 0.85 | 0.82 | 0.80 | 0.77 | 0.74 | 0.72 | 0.70 | 0.67 | 0.63 |
| | Tendu | LS/ha | 1 | | | 80.00 | 80.00 | 80.00 | 60.00 | 60.00 | 60.00 | 60.00 | 60.00 | 60.00 | 60.00 |
| | NTFP | | 1 | | | 50.00 | 50.00 | 50.00 | 150.00 | 150.00 | 150.00 | 150.00 | 150.00 | 150.00 | 150.00 |

**Summary of benefits without JFM**

| | | Net Worth (Rs) | Discount Rate (%) | | 0 | 1 | 2 | 6 | 10 | 14 | 18 | 22 | 25 | 30 | 35 |
|---|---|---|---|---|---|---|---|---|---|---|---|---|---|---|---|
| FD | | 18,240 | 12 | | | 1,974 | 1,974 | 1,974 | 1,974 | 1,974 | 1,974 | 1,974 | 1,974 | 1,974 | 112,946 |
| FPC | Revenue | 0 | 12 | | | 0 | 0 | 0 | 0 | 0 | 0 | 0 | 0 | 0 | 0 |
| | Headloaders | 2,629 | 12 | | | 364 | 333 | 325 | 315 | 305 | 296 | 287 | 280 | 269 | 259 |
| | Graziers | 2,703 | 12 | | | 350 | 350 | 338 | 327 | 316 | 305 | 294 | 286 | 273 | 260 |
| | Collectors | 1,291 | 12 | | | 130 | 130 | 130 | 210 | 210 | 210 | 210 | 210 | 210 | 210 |
| | Total | 6,622 | 12 | | | 844 | 813 | 793 | 852 | 831 | 811 | 791 | 776 | 752 | 729 |
| Economic | | 24,862 | 12 | | | 2,818 | 2,787 | 2,767 | 2,826 | 2,805 | 2,785 | 2,765 | 2,750 | 2,726 | 113,675 |

**INCREMENTAL BENEFITS DUE TO JFM**

| | | Net Worth (Rs) | Discount Rate (%) | | 0 | 1 | 2 | 6 | 10 | 14 | 18 | 22 | 25 | 30 | 35 |
|---|---|---|---|---|---|---|---|---|---|---|---|---|---|---|---|
| FD | | 14,238 | 12 | | | -1,974 | 12,349 | 12,349 | 12,349 | 12,349 | 12,349 | 12,349 | -1,974 | 12,349 | -24,235 |
| FPC | Revenue | 32,478 | 12 | | | 0 | 14,323 | 14,323 | 14,323 | 14,323 | 14,323 | 14,323 | 0 | 14,323 | 88,712 |
| | Headloaders | -2,423 | 12 | | | -324 | -296 | -298 | -295 | -291 | -285 | -279 | -274 | -265 | -256 |
| | Graziers | -984 | 12 | | | 43 | 12 | -92 | -176 | -242 | -290 | -324 | -342 | -362 | -372 |
| | Collectors | 1,279 | 12 | | | 0 | 80 | 160 | 180 | 280 | 280 | 280 | 280 | 280 | 230 |
| | Total | 30,349 | 12 | | | -281 | 14,119 | 14,094 | 14,032 | 14,071 | 14,028 | 14,000 | -336 | 13,976 | 88,314 |
| Economic | | 44,587 | 12 | | | -2,255 | 26,468 | 26,443 | 26,381 | 26,420 | 26,378 | 26,350 | -2,310 | 26,325 | 64,079 |

**Table MT.7**  **Forest Productivity Model (Type NF-4)**

| Management | Output | Unit | Unit Price (Rs) | Offtake rates (% CAI) | Year 0 | 1 | 2 | 6 | 10 | 14 | 18 | 22 | 25 | 30 | 35 |
|---|---|---|---|---|---|---|---|---|---|---|---|---|---|---|---|
| **WITH JFM** | | | | | | | | | | | | | | | |
| | GS | M^3/ha | | | 18.91 | 22.63 | 25.68 | 35.73 | 43.08 | 48.46 | 52.40 | 55.28 | 56.92 | 58.93 | 60.29 |
| | CAI | M^3/ha | | | 3.73 | 3.41 | 3.16 | 2.31 | 1.69 | 1.24 | 0.90 | 0.66 | 0.52 | 0.35 | 0.24 |
| | CC | (%) | | | 14 | 18 | 21 | 35 | 47 | 56 | 64 | 70 | 73 | 78 | 81 |
| | Dead/dry | M^3/ha | 135.00 | 10 % | | 0.37 | 0.34 | 0.25 | 0.18 | 0.13 | 0.10 | 0.07 | 0.06 | 0.04 | 0.03 |
| | Green cut | M^3/ha | 135.00 | 0 % | | 0.00 | 0.00 | 0.00 | 0.00 | 0.00 | 0.00 | 0.00 | 0.00 | 0.00 | 0.00 |
| | Teak | M^3/ha | 8,324.00 | | | | | | | | | | | | 12.06 |
| | Fuelwood | M^3/ha | 1,152.00 | | | | | | | | | | | | 3.01 |
| | Non teak | M^3/ha | 459.00 | | | | | | | | | | | | 31.65 |
| | Fuelwood | M^3/ha | 135.00 | | | | | | | | | | | | 13.56 |
| | Bamboo | Poles | 8.00 | | 400 | | 3,200 | 3,200 | 3,200 | 3,200 | 3,200 | 3,200 | | 3,200 | |
| | Lops/tops | kg/pole | 0.11 | | 8.50 | | 27,200 | 27,200 | 27,200 | 27,200 | 27,200 | 27,200 | | 27,200 | |
| | Grazing | MT/ha | 460.00 | | | 1.05 | 0.98 | 0.71 | 0.47 | 0.27 | 0.12 | 0.00 | -0.07 | -0.16 | -0.22 |
| | Tendu | LS/ha | 1 | | | 100.00 | 80.00 | 60.00 | 40.00 | 40.00 | 40.00 | 40.00 | 40.00 | 40.00 | 40.00 |
| | NTFP | | 1 | | | 25.00 | 50.00 | 150.00 | 250.00 | 350.00 | 450.00 | 450.00 | 450.00 | 450.00 | 400.00 |

**Summary of benefits with JFM**

| | | Net Worth (Rs) | Discount Rate (%) | | 0 | 1 | 2 | 6 | 10 | 14 | 18 | 22 | 25 | 30 | 35 |
|---|---|---|---|---|---|---|---|---|---|---|---|---|---|---|---|
| FD | | 31,936 | 12 | | | 0 | 14,323 | 14,323 | 14,323 | 14,323 | 14,323 | 14,323 | 0 | 14,323 | 60,097 |
| FPC | Revenue | 31,936 | 12 | | | 0 | 14,323 | 14,323 | 14,323 | 14,323 | 14,323 | 14,323 | 0 | 14,323 | 60,097 |
| | Headloaders | 255 | 12 | | | 50 | 46 | 34 | 25 | 18 | 13 | 10 | 8 | 5 | 3 |
| | Graziers | 2,287 | 12 | | | 483 | 450 | 325 | 215 | 125 | 55 | 0 | -32 | -72 | -100 |
| | Collectors | 2,117 | 12 | | | 125 | 130 | 210 | 290 | 390 | 490 | 490 | 490 | 490 | 440 |
| | Total | 36,595 | 12 | | | 658 | 14,949 | 14,892 | 14,853 | 14,857 | 14,881 | 14,823 | 466 | 14,746 | 60,440 |
| Economic | | 68,531 | 12 | | | 658 | 29,273 | 29,215 | 29,176 | 29,180 | 29,204 | 29,146 | 466 | 29,069 | 120,537 |

| Management | Output | Unit | Unit Price (Rs) | Offtake rates (% CAI) | Year 0 | 1 | 2 | 6 | 10 | 14 | 18 | 22 | 25 | 30 | 35 |
|---|---|---|---|---|---|---|---|---|---|---|---|---|---|---|---|
| **WITHOUT JFM** | | | | | | | | | | | | | | | |
| | GS | M^3/ha | | | 18.91 | 22.63 | 21.58 | 19.01 | 16.24 | 13.30 | 10.17 | 6.85 | 4.22 | 0.00 | 0.00 |
| | CAI | M^3/ha | | | 3.73 | 3.41 | 3.50 | 3.72 | 3.95 | 4.20 | 4.46 | 4.74 | 4.97 | 0.00 | 0.00 |
| | CC | (%) | | | 14 | 18 | 16 | 14 | 11 | 9 | 6 | 4 | 2 | 0 | 0 |
| | Dead/dry | M^3/ha | 135.00 | 10 % | | 0.37 | 0.34 | 0.37 | 0.39 | 0.41 | 0.44 | 0.47 | 0.49 | 0.00 | 0.00 |
| | Green cut | M^3/ha | 135.00 | 110 % | | 4.10 | 3.75 | 4.03 | 4.28 | 4.55 | 4.84 | 5.14 | 5.38 | 0.00 | 0.00 |
| | Teak | M^3/ha | 8,324.00 | | | | | | | | | | | | 0.00 |
| | Fuelwood | M^3/ha | 1,152.00 | | | | | | | | | | | | 0.00 |
| | Non teak | M^3/ha | 459.00 | | | | | | | | | | | | 0.00 |
| | Fuelwood | M^3/ha | 135.00 | | | | | | | | | | | | 0.00 |
| | Bamboo | Poles | 2.00 | | 400 | 800 | 800 | 800 | 800 | 800 | 800 | 800 | 800 | 800 | 800 |
| | Lops/tops | kg/pole | 0.11 | | 4.25 | 3,400 | 3,400 | 3,400 | 3,400 | 3,400 | 3,400 | 3,400 | 3,400 | 3,400 | 3,400 |
| | Grazing | MT/ha | 410.00 | | | 1.05 | 1.07 | 1.13 | 1.18 | 1.23 | 1.28 | 1.32 | 1.36 | 1.40 | 1.40 |
| | Tendu | LS/ha | 1 | | | 100.00 | 100.00 | 100.00 | 100.00 | 120.00 | 120.00 | 120.00 | 120.00 | 120.00 | 120.00 |
| | NTFP | | 1 | | | 25.00 | 25.00 | 25.00 | 25.00 | 0.00 | 0.00 | 0.00 | 0.00 | 0.00 | 0.00 |

**Summary of benefits without JFM**

| | | Net Worth (Rs) | Discount Rate (%) | | 0 | 1 | 2 | 6 | 10 | 14 | 18 | 22 | 25 | 30 | 35 |
|---|---|---|---|---|---|---|---|---|---|---|---|---|---|---|---|
| FD | | 16,138 | 12 | | | 1,974 | 1,974 | 1,974 | 1,974 | 1,974 | 1,974 | 1,974 | 1,974 | 1,974 | 1,974 |
| FPC | Revenue | 0 | 12 | | | 0 | 0 | 0 | 0 | 0 | 0 | 0 | 0 | 0 | 0 |
| | Headloaders | 4,961 | 12 | | | 604 | 553 | 593 | 631 | 670 | 712 | 757 | 792 | 0 | 0 |
| | Graziers | 3,875 | 12 | | | 430 | 440 | 461 | 483 | 504 | 524 | 543 | 556 | 572 | 572 |
| | Collectors | 1,011 | 12 | | | 125 | 125 | 125 | 125 | 120 | 120 | 120 | 120 | 120 | 120 |
| | Total | 9,847 | 12 | | | 1,159 | 1,118 | 1,180 | 1,239 | 1,294 | 1,356 | 1,420 | 1,468 | 692 | 692 |
| Economic | | 25,985 | 12 | | | 3,133 | 3,092 | 3,154 | 3,213 | 3,268 | 3,330 | 3,394 | 3,442 | 2,666 | 2,666 |

**INCREMENTAL BENEFITS DUE TO JFM**

| | | Net Worth (Rs) | Discount Rate (%) | | 0 | 1 | 2 | 6 | 10 | 14 | 18 | 22 | 25 | 30 | 35 |
|---|---|---|---|---|---|---|---|---|---|---|---|---|---|---|---|
| FD | | 15,798 | 12 | | | -1,974 | 12,349 | 12,349 | 12,349 | 12,349 | 12,349 | 12,349 | -1,974 | 12,349 | 58,123 |
| FPC | Revenue | 31,936 | 12 | | | 0 | 14,323 | 14,323 | 14,323 | 14,323 | 14,323 | 14,323 | 0 | 14,323 | 60,097 |
| | Headloaders | -4,705 | 12 | | | -553 | -507 | -560 | -606 | -652 | -699 | -747 | -785 | 5 | 3 |
| | Graziers | -1,588 | 12 | | | 52 | 11 | -137 | -268 | -379 | -470 | -543 | -587 | -645 | -673 |
| | Collectors | 1,106 | 12 | | | 0 | 5 | 85 | 165 | 270 | 370 | 370 | 370 | 370 | 320 |
| | Total | 26,748 | 12 | | | -501 | 13,832 | 13,712 | 13,614 | 13,562 | 13,525 | 13,403 | -1,002 | 14,054 | 59,748 |
| Economic | | 42,546 | 12 | | | -2,475 | 26,181 | 26,061 | 25,964 | 25,912 | 25,874 | 25,753 | -2,976 | 26,403 | 117,871 |

**Table MT.8    Forest productivity model (Plantations)**

| Plantation costs | 1 | 2 | 3 | 4 | 5 | 6 | 7 | 8 | 9 | 10 |
|---|---|---|---|---|---|---|---|---|---|---|
| Labour | 404 | 7,221 | 19,971 | 83,923 | 62,962 | 146,923 | 51,115 | 17,462 | 3,692 | 3,077 |
| Materials | 725 | 1,850 | 10,100 | 8,900 | 23,300 | 10,400 | 12,000 | -0 | 0 | 0 |
| Total | 1,775 | 20,625 | 62,025 | 227,100 | 187,000 | 392,400 | 144,900 | 45,400 | 9,600 | 8,000 |

**Table MT.9  (A) Years 1-15   Forest Productivity Model (Plantation Outputs)**

| Plantation | Output | Unit | Area ha | MAI | Rotation | Stems No/ha | Price per unit | 1 | 2 | 3 | 4 | 5 | 6 | 7 | 8 | 9 | 10 | 11 | 12 | 13 | 14 | 15 |
|---|---|---|---|---|---|---|---|---|---|---|---|---|---|---|---|---|---|---|---|---|---|---|
| PL-1 Teak | Timber | m^3 | 2.5 | 15 | 50 | 1800 | | | | | | | | | | | | | | | | |
| | Teak Poles (11) | poles | | | | | | | | | | | | | | | | 2250 | | | | |
| | Teak poles (25) | poles | | | | | | | | | | | | | | | | | | | | |
| | Fuelwood | m^3 | | | | | | | | | | | | | | | | | | | | |
| | Bamboo | poles | | 4 | | 100 | | | 500 | | | 750 | | | | 1500 | | | | 2000 | | |
| | Lops and tops | kg | | | | 8.5 | | | 4250 | | | 6375 | | | | 12750 | | | | 17000 | | |
| | Grazing | | | | | | | 3625 | 3000 | 3000 | 2750 | 1875 | 1625 | 1625 | 1625 | 1250 | 1000 | 1000 | 1000 | 1000 | 1000 | 1000 |
| PL-2 Sevan | Timber | m^3 | 5.0 | 10.2 | 50 | 300 | | | | | | | | | | | | | | | | |
| | Fuelwood | m^3 | | | | | | | | | | | | | | | | | | | | |
| | Grazing | | | | | | | 7250 | 6000 | 6000 | 5500 | 3750 | 3250 | 3250 | 3250 | 2500 | 2000 | 2000 | 2000 | 2000 | 2000 | 2000 |
| PL-3 SWC/MF fodder | Teak timber | m^3 | 30.0 | 15 | 50 | 100 | | | | | | | | | | | | | | | | |
| | Teak Poles (11) | poles | | | | | | | | | | | | | | | | | | | | |
| | Teak poles (25) | poles | | | | | | | | | | | | | | | | | | | | |
| | Fuelwood | m^3 | | | | | | | | | | | | | 1500 | | | | | | | |
| | Non-teak fuelwo | m^3 | | 3 | 30 | 320 | | | | | | | 28.8 | 28.8 | 28.8 | 28.8 | 28.8 | 28.8 | 28.8 | 28.8 | 28.8 | 28.8 |
| | Bamboo | poles | | | | 150 | | | | 13500 | | | 27000 | | | | 36000 | | | | 36000 | |
| | Lops and tops | m^3 | | | | 8.5 | | | | 114750 | | | 229500 | | | | 306000 | | | | 306000 | |
| | Fruit | kg | | | | 185 | | | | | | 5550 | 5550 | 5550 | 5550 | 55500 | 55500 | 55500 | 55500 | 55500 | 111000 | 111000 |
| | Fuelwood | m^3 | | | | | | | | | | | | | | | | | | | | |
| | Grazing | | | | | | | 43500 | 36000 | 36000 | 33000 | 22500 | 19500 | 19500 | 19500 | 15000 | 12000 | 12000 | 12000 | 12000 | 12000 | 12000 |
| PL-4 Fodder | Stylos | MT | 10.0 | 5 | 5 | | | 50 | 50 | 50 | 50 | 50 | 50 | 50 | 50 | 50 | 50 | 50 | 50 | 50 | 50 | 50 |
| PL-5 Wasteland mixed | Non-teak timber | | 50.0 | 5.2 | 30 | 490 | | | | | | | | 127.4 | 127.4 | 127.4 | 127.4 | 127.4 | 127.4 | 127.4 | 127.4 | 127.4 |
| | Bamboo | | | | | 100 | | | 10000 | | | | 15000 | | | | 30000 | | | | 40000 | |
| | Lops and tops | | | | | 8.5 | | | 85000 | | | | 127500 | | | | 255000 | | | | 340000 | |
| | Fruit/Medicinal plants | | | | 30 | 50 | | | | | | | 2500 | 2500 | 2500 | 2500 | 2500 | 25000 | 25000 | 25000 | 25000 | 50000 |
| | Fuelwood | m^3 | | | | | | | | | | | | | | | | | | | | |
| | Grazing | | | | | | | 72500 | 60000 | 60000 | 55000 | 37500 | 32500 | 32500 | 32500 | 25000 | 20000 | 20000 | 20000 | 20000 | 20000 | 20000 |
| **Total Plantation Outputs** | | | | | | | | | | | | | | | | | | | | | | |
| Teak | Timber | m^3 | | | | | 8,324.0 | 0 | 0 | 0 | 0 | 0 | 0 | 0 | 0 | 0 | 0 | 0 | 0 | 0 | 0 | 0 |
| | Medium poles | pole | | | | | 60.5 | 0 | 0 | 0 | 0 | 0 | 0 | 0 | 0 | 0 | 0 | 2250 | 0 | 0 | 0 | 0 |
| | Small poles | pole | | | | | 40.3 | 0 | 0 | 0 | 0 | 0 | 0 | 0 | 0 | 0 | 0 | 0 | 0 | 0 | 0 | 0 |
| | Fuelwood | m^3 | | | | | 1,152.0 | 0 | 0 | 0 | 0 | 0 | 0 | 0 | 0 | 0 | 0 | 0 | 0 | 0 | 0 | 0 |
| Sewan | Timber | m^3 | | | | | 5,000.0 | 0 | 0 | 0 | 0 | 0 | 0 | 0 | 0 | 0 | 0 | 0 | 0 | 0 | 0 | 0 |
| | Fuelwood | m^3 | | | | | 132.0 | 0 | 0 | 0 | 0 | 0 | 0 | 0 | 0 | 0 | 0 | 0 | 0 | 0 | 0 | 0 |
| Other | Fuelwood | m^3 | | | | | 132.0 | 0 | 0 | 0 | 0 | 0 | 28.8 | 156.2 | 156.2 | 156.2 | 156.2 | 156.2 | 156.2 | 156.2 | 156.2 | 156.2 |
| Bamboo | Poles | Poles | | | | | 8.0 | 0 | 10500 | 13500 | 0 | 750 | 42000 | 0 | 0 | 1500 | 66000 | 0 | 0 | 2000 | 76000 | 0 |
| | Lops and tops | Kg | | | | | 0.1 | 0 | 89250 | 114750 | 0 | 6375 | 357000 | 0 | 0 | 12750 | 561000 | 0 | 0 | 17000 | 646000 | 0 |
| MFP | Fruit/Medicinal plants | | | | | | 2.5 | 0 | 0 | 0 | 0 | 5550 | 8050 | 8050 | 8050 | 58000 | 58000 | 80500 | 80500 | 80500 | 136000 | 161000 |
| | Fuelwood | | | | | | 132 | 0 | 0 | 0 | 0 | 0 | 0 | 0 | 0 | 0 | 0 | 0 | 0 | 0 | 0 | 0 |
| Stylo | | MT | | | | | 492 | 50 | 50 | 50 | 50 | 50 | 50 | 50 | 50 | 50 | 50 | 50 | 50 | 50 | 50 | 50 |
| Grazing | | MT | | | | | 0.5 | 126875 | 105000 | 105000 | 96250 | 65625 | 56875 | 56875 | 56875 | 43750 | 35000 | 35000 | 35000 | 35000 | 35000 | 35000 |

**Table MT.9  (B) Years 16–34    Forest Productivity Model (Plantation Outputs)**

| Plantation | Output | Unit | 16 | 17 | 18 | 19 | 20 | 21 | 22 | 23 | 24 | 25 | 26 | 27 | 28 | 29 | 30 | 31 | 32 | 33 | 34 |
|---|---|---|---|---|---|---|---|---|---|---|---|---|---|---|---|---|---|---|---|---|---|
| **PL-1 Teak** | Timber | m^3 | | | | | | | | | | | | | | | | | | | |
| | Teak Poles (11) | poles | | | | | | | | | | | | | | | | | | | |
| | Teak poles (25) | poles | | | | | | | | | | 1125 | | | | | | | | | |
| | Fuelwood | m^3 | | | | | | | | | | | | | | | | | | | |
| | Bamboo | poles | | 2000 | | | | 2000 | | | | 2000 | | | | 2000 | | | | 2000 | |
| | Lops and tops | kg | | 17000 | | | | 17000 | | | | 17000 | | | | 17000 | | | | 17000 | |
| | Grazing | kg | 0 | | | | | | | | | | | | | | | | | | |
| **PL-2 Sevan** | Timber | m^3 | | | | | | | | | | | | | | | | | | | |
| | Fuelwood | m^3 | | | | | | | | | | | | | | | | | | | |
| | Grazing | | 0 | | | | | | | | | | | | | | | | | | |
| **PL-3 SW/C/MFP/ fodder** | Teak timber | m^3 | | | | | | | | | | | | | | | | | | | |
| | Teak Poles (11) | poles | | | | | | | | | | | | | | | | | | | |
| | Teak poles (25) | poles | | | | | | | 750 | | | | | | | | | | | | |
| | Fuelwood | m^3 | | | | | | | | | | | | | | | | | | | |
| | Non-teak fuelwood | m^3 | 28.8 | 28.8 | 28.8 | 28.8 | 28.8 | 28.8 | 28.8 | 28.8 | 28.8 | 28.8 | 28.8 | 28.8 | 28.8 | 28.8 | 28.8 | 28.8 | 28.8 | 28.8 | 28.8 |
| | Bamboo | poles | | | 36000 | | | | 36000 | | | | 36000 | | | | 36000 | | | | 36000 |
| | Lops and tops | m^3 | | | 306000 | | | | 306000 | | | | 306000 | | | | 306000 | | | | 306000 |
| | Fruit | kg | 111000 | 111000 | 111000 | 166500 | 166500 | 166500 | 166500 | 166500 | 222000 | 222000 | 222000 | | | | | | | | |
| | Fuelwood | m^3 | | | | | | | | | | | 5.55 | 2.5 | | | | | | | |
| | Grazing | | 0 | | | | | | | | | | | | | | | | | | |
| **PL-4 Fodder** | Stylos | MT | 50 | 50 | 50 | 50 | 50 | 50 | 50 | 50 | 50 | 50 | 50 | 50 | 50 | 50 | 50 | 50 | 50 | 50 | 50 |
| | Grazing | | 0 | | | | | | | | | | | | | | | | | | |
| **PL-5 Wasteland/ mixed** | Non-teak timber | m^3 | 127.4 | 127.4 | 127.4 | 127.4 | 127.4 | 127.4 | 127.4 | 127.4 | 127.4 | 127.4 | 127.4 | 127.4 | 127.4 | 127.4 | 127.4 | 127.4 | 127.4 | 127.4 | 127.4 |
| | Bamboo | poles | | | 40000 | | | | 40000 | | | | 40000 | | | | 40000 | | | | 40000 |
| | Lops and tops | pole | | | 340000 | | | | 340000 | | | | 340000 | | | | 340000 | | | | 340000 |
| | Fuelwood | m^3 | 50000 | 50000 | 50000 | 50000 | 75000 | 75000 | 75000 | 75000 | 75000 | 100000 | 100000 | 100000 | | | | | | | |
| | Fruit/Medicinal plants | m^3 | | | | | | | | | | 100000 | 100000 | 100000 | | | | | | | |
| | Fuelwood | m^3 | | | | | | | | | | | | | | | | | | | |
| | Grazing | m^3 | | | | | | | | | | | 5.55 | 2.5 | | | | | | | |
| **Total Plantation Outputs** | | | | | | | | | | | | | | | | | | | | | |
| **Teak** | Timber | m^3 | 0 | 0 | 0 | 0 | 0 | 0 | 0 | 0 | 0 | 0 | 0 | 0 | 0 | 0 | 0 | 0 | 0 | 0 | 0 |
| | Medium poles | pole | 0 | 0 | 0 | 0 | 0 | 0 | 0 | 0 | 0 | 0 | 0 | 0 | 0 | 0 | 0 | 0 | 0 | 0 | 0 |
| | Small poles | pole | 0 | 0 | 0 | 0 | 0 | 0 | 750 | 0 | 0 | 1125 | 0 | 0 | 0 | 0 | 0 | 0 | 0 | 0 | 0 |
| | Fuelwood | m^3 | 0 | 0 | 0 | 0 | 0 | 0 | 0 | 0 | 0 | 0 | 0 | 0 | 0 | 0 | 0 | 0 | 0 | 0 | 0 |
| **Sewan** | Timber | m^3 | 0 | 0 | 0 | 0 | 0 | 0 | 0 | 0 | 0 | 0 | 0 | 0 | 0 | 0 | 0 | 0 | 0 | 0 | 0 |
| | Fuelwood | m^3 | 0 | 0 | 0 | 0 | 0 | 0 | 0 | 0 | 0 | 0 | 0 | 0 | 0 | 0 | 0 | 0 | 0 | 0 | 0 |
| **Other** | Fuelwood | m^3 | 156.2 | 156.2 | 156.2 | 156.2 | 156.2 | 156.2 | 156.2 | 156.2 | 156.2 | 156.2 | 156.2 | 156.2 | 156.2 | 156.2 | 156.2 | 156.2 | 156.2 | 156.2 | 156.2 |
| **Bamboo** | Poles | Poles | 0 | 2000 | 76000 | 0 | 0 | 2000 | 76000 | 0 | 0 | 2000 | 76000 | 0 | 0 | 2000 | 76000 | 0 | 0 | 2000 | 76000 |
| | Lops and tops | Kg | 0 | 17000 | 646000 | 0 | 0 | 17000 | 646000 | 0 | 0 | 17000 | 646000 | 0 | 0 | 17000 | 646000 | 0 | 0 | 17000 | 646000 |
| **MFP** | Fruit/ Medicinal plants | | 161000 | 161000 | 161000 | 216500 | 241500 | 241500 | 241500 | 241500 | 297000 | 322000 | 322000 | 100000 | 0 | 0 | 0 | 0 | 0 | 0 | 0 |
| | Fuelwood | | 0 | 0 | 0 | 0 | 0 | 0 | 0 | 0 | 0 | 0 | 5.55 | 2.5 | 0 | 0 | 0 | 0 | 0 | 0 | 0 |
| **Stylo** | | MT | 50 | 50 | 50 | 50 | 50 | 50 | 50 | 50 | 50 | 50 | 50 | 50 | 50 | 50 | 50 | 50 | 50 | 50 | 50 |
| **Grazing** | | MT | 0 | 0 | 0 | 0 | 0 | 0 | 0 | 0 | 0 | 0 | 0 | 0 | 0 | 0 | 0 | 0 | 0 | 0 | 0 |

**Table MT.9    (C) Years 35-50    Forest Productivity Model (Plantation Outputs)**

| Plantation | Output | Unit | 35 | 36 | 37 | 38 | 39 | 40 | 41 | 42 | 43 | 44 | 45 | 46 | 47 | 48 | 49 | 50 |
|---|---|---|---|---|---|---|---|---|---|---|---|---|---|---|---|---|---|---|
| PL-1 Teak | Timber | m^3 | | | | | | | | | | | | | | | | 30 |
| | Teak Poles (11) | poles | | | | | | | | | | | | | | | | |
| | Teak poles (25) | poles | | | | | | | | | | | | | | | | |
| | Fuelwood | m^3 | | | | | | | | | | | | | | | | 7.5 |
| | Bamboo | poles | | | 2000 | | | | 2000 | | | | 2000 | | | | 2000 | |
| | Lops and tops | kg | | | 17000 | | | | 17000 | | | | 17000 | | | | 17000 | |
| | Grazing | | | | | | | | | | | | | | | | | |
| PL-2 Sevan | Timber | m^3 | | | | | | | | | | | | | | | | |
| | Fuelwood | m^3 | | | | | | | | | | | | 10.71 | | | | |
| | Grazing | m^3 | | | | | | | | | | | | 4.59 | | | | |
| PL-3 SWC/MFP/fodder | Teak timber | m^3 | | | | | | | | | | | | | | | | |
| | Teak Poles (11) | poles | | | | | | | | | | | | | 36 | | | |
| | Teak poles (25) | poles | | | | | | | | | | | | | 9 | | | |
| | Fuelwood | m^3 | | | | | | | | | | | | | | | | |
| | Non-teak fuelwood | m^3 | 28.8 | 28.8 | 28.8 | 28.8 | 28.8 | 28.8 | 28.8 | 28.8 | 28.8 | 28.8 | 28.8 | 28.8 | 28.8 | 28.8 | 28.8 | 28.8 |
| | Bamboo | poles | | | | 36000 | | | | 36000 | | | | 36000 | | | | 36000 |
| | Lops and tops | m^3 | | | | 306000 | | | | 306000 | | | | 306000 | | | | 306000 |
| | Fruit | kg | 0 | 0 | 5550 | 5550 | 5550 | 5550 | 55500 | 55500 | 55500 | 55500 | 55500 | 111000 | 111000 | 111000 | 111000 | 111000 |
| | Fuelwood | m^3 | 0 | 0 | 5550 | 5550 | 5550 | 5550 | 55500 | 55500 | 55500 | 55500 | 55500 | 111000 | 111000 | 111000 | 111000 | 111000 |
| | Grazing | | | | | | | | | | | | | | | | | |
| PL-4 Fodder | Stylos | MT | 50 | 50 | 50 | 50 | 50 | 50 | 50 | 50 | 50 | 50 | 50 | 50 | 50 | 50 | 50 | 50 |
| PL-5 Wasteland/mixed | Non-teak timber | m^3 | 127.4 | 127.4 | 127.4 | 127.4 | 127.4 | 127.4 | 127.4 | 127.4 | 127.4 | 127.4 | 127.4 | 127.4 | 127.4 | 127.4 | 127.4 | 127.4 |
| | Bamboo | poles | | | | 40000 | | | | 40000 | | | | 40000 | | | | 40000 |
| | Lops and tops | | | | | 340000 | | | | 340000 | | | | 340000 | | | | 340000 |
| | Fuelwood | m^3 | | | | | | | | | | | | | | | | |
| | Fruit/Medicinal plants | | 0 | 0 | 0 | 2500 | 2500 | 2500 | 2500 | 2500 | 25000 | 25000 | 25000 | 25000 | 25000 | 50000 | 50000 | 50000 |
| | Fuelwood | | | | | | | | | | | | | | | | | |
| | Grazing | | | | | | | | | | | | | | | | | |

**Total Plantation Outputs**

| Plantation | Output | Unit | 35 | 36 | 37 | 38 | 39 | 40 | 41 | 42 | 43 | 44 | 45 | 46 | 47 | 48 | 49 | 50 |
|---|---|---|---|---|---|---|---|---|---|---|---|---|---|---|---|---|---|---|
| Teak | Timber | m^3 | 0 | 0 | 0 | 0 | 0 | 0 | 0 | 0 | 0 | 0 | 0 | 0 | 0 | 0 | 0 | 30 |
| | Medium poles | pole | 0 | 0 | 0 | 0 | 0 | 0 | 0 | 0 | 0 | 0 | 0 | 0 | 36 | 0 | 0 | 0 |
| | Small poles | pole | 0 | 0 | 0 | 0 | 0 | 0 | 0 | 0 | 0 | 0 | 0 | 0 | 9 | 0 | 0 | 0 |
| | Fuelwood | m^3 | 0 | 0 | 0 | 0 | 0 | 0 | 0 | 0 | 0 | 0 | 0 | 0 | 0 | 0 | 0 | 7.5 |
| Sewan | Timber | m^3 | 0 | 0 | 0 | 0 | 0 | 0 | 0 | 0 | 0 | 0 | 0 | 10.71 | 0 | 0 | 0 | 0 |
| | Fuelwood | m^3 | 0 | 0 | 0 | 0 | 0 | 0 | 0 | 0 | 0 | 0 | 0 | 4.59 | 0 | 0 | 0 | 0 |
| Other | Fuelwood | m^3 | 156.2 | 156.2 | 156.2 | 156.2 | 156.2 | 156.2 | 156.2 | 156.2 | 156.2 | 156.2 | 156.2 | 156.2 | 156.2 | 156.2 | 156.2 | 156.2 |
| Bamboo | Poles | Poles | 0 | 0 | 2000 | 76000 | 0 | 0 | 2000 | 76000 | 0 | 0 | 2000 | 76000 | 0 | 0 | 2000 | 76000 |
| | Lops and tops | Kg | 0 | 0 | 17000 | 646000 | 0 | 0 | 17000 | 646000 | 0 | 0 | 17000 | 646000 | 0 | 0 | 17000 | 646000 |
| MFP | Fruit/ Medicinal plants | | 0 | 0 | 5550 | 8050 | 8050 | 8050 | 58000 | 58000 | 80500 | 80500 | 80500 | 136000 | 136000 | 161000 | 161000 | 161000 |
| | Fuelwood | | 0 | 0 | 0 | 0 | 0 | 0 | 0 | 0 | 0 | 0 | 0 | 0 | 0 | 0 | 0 | 0 |
| Stylo | | MT | 50 | 50 | 50 | 50 | 50 | 50 | 50 | 50 | 50 | 50 | 50 | 50 | 50 | 50 | 50 | 50 |
| Grazing | | MT | 0 | 0 | 0 | 0 | 0 | 0 | 0 | 0 | 0 | 0 | 0 | 0 | 0 | 0 | 0 | 0 |

**Table MT.10 (A) Years 1-15          Forest Productivity Model (Value of Total Plantation Outputs)**

| Plantation | Output | Unit | 1 | 2 | 3 | 4 | 5 | 6 | 7 | 8 | 9 | 10 | 11 | 12 | 13 | 14 | 15 |
|---|---|---|---|---|---|---|---|---|---|---|---|---|---|---|---|---|---|
| | | | | | | | | | | | | | | | | Year | |
| Teak | Timber | m^3 | 0 | 0 | 0 | 0 | 0 | 0 | 0 | 0 | 0 | 0 | 0 | 0 | 0 | 0 | 0 |
| | Medium poles | pole | 0 | 0 | 0 | 0 | 0 | 0 | 0 | 90,765 | 0 | 0 | 136,148 | 0 | 0 | 0 | 0 |
| | Small poles | pole | 0 | 0 | 0 | 0 | 0 | 0 | 0 | 0 | 0 | 0 | 0 | 0 | 0 | 0 | 0 |
| | Fuelwood | m^3 | 0 | 0 | 0 | 0 | 0 | 0 | 0 | 0 | 0 | 0 | 0 | 0 | 0 | 0 | 0 |
| Sewan | Timber | m^3 | 0 | 0 | 0 | 0 | 0 | 0 | 0 | 0 | 0 | 0 | 0 | 0 | 0 | 0 | 0 |
| | Fuelwood | m^3 | 0 | 0 | 0 | 0 | 0 | 0 | 0 | 0 | 0 | 0 | 0 | 0 | 0 | 0 | 0 |
| Non-teak | Fuelwood | m^3 | 0 | 0 | 0 | 0 | 0 | 3,802 | 20,618 | 20,618 | 20,618 | 20,618 | 20,618 | 20,618 | 20,618 | 20,618 | 20,618 |
| Bamboo | Poles | Poles | 0 | 84,000 | 108,000 | 0 | 6,000 | 336,000 | 0 | 0 | 12,000 | 528,000 | 0 | 0 | 16,000 | 608,000 | 0 |
| | Lops and tops | Kg | 0 | 9,996 | 12,852 | 0 | 714 | 39,984 | 0 | 0 | 1,428 | 62,832 | 0 | 0 | 1,904 | 72,352 | 0 |
| MFP | Fruit/Medicinal plants | | 0 | 0 | 0 | 0 | 13,875 | 20,125 | 20,125 | 20,125 | 145,000 | 145,000 | 201,250 | 201,250 | 201,250 | 340,000 | 402,500 |
| | Fuelwood | m^3 | 0 | 0 | 0 | 0 | 0 | 0 | 0 | 0 | 0 | 0 | 0 | 0 | 0 | 0 | 0 |
| Stylo | | | 0 | 24,600 | 24,600 | 24,600 | 24,600 | 24,600 | 24,600 | 24,600 | 24,600 | 24,600 | 24,600 | 24,600 | 24,600 | 24,600 | 24,600 |
| Grazing | | MT | 62,423 | 51,660 | 51,660 | 47,355 | 32,288 | 27,983 | 27,983 | 27,983 | 21,525 | 17,220 | 17,220 | 17,220 | 17,220 | 17,220 | 17,220 |
| Total | | | 62,423 | 170,256 | 197,112 | 71,955 | 77,477 | 452,493 | 93,326 | 184,091 | 225,171 | 798,270 | 399,836 | 263,688 | 281,592 | 1,082,790 | 464,938 |

**Distribution of Benefit Flows From Plantations Under JFM**

| | Net Worth (Rs) | Discount Rate (%) | 1 | 2 | 3 | 4 | 5 | 6 | 7 | 8 | 9 | 10 | 11 | 12 | 13 | 14 | 15 |
|---|---|---|---|---|---|---|---|---|---|---|---|---|---|---|---|---|---|
| FD Revenue | 511,479 | 12 | 0 | 46,998 | 60,426 | 0 | 3,357 | 187,992 | 0 | 45,383 | 6,714 | 295,416 | 68,074 | 0 | 8,952 | 340,176 | 0 |
| - Costs | 601,331 | 12 | 1,775 | 20,625 | 62,025 | 227,100 | 187,000 | 392,400 | 144,900 | 45,400 | 9,600 | 8,000 | 0 | 0 | 0 | 0 | 0 |
| Net revenue | -89,853 | 12 | -1,775 | 26,373 | -1,599 | -227,100 | -183,643 | -204,408 | -144,900 | -18 | -2,886 | 287,416 | 68,074 | 0 | 8,952 | 340,176 | 0 |
| Vill Revenue | 511,479 | 12 | 0 | 46,998 | 60,426 | 0 | 3,357 | 187,992 | 0 | 45,383 | 6,714 | 295,416 | 68,074 | 0 | 8,952 | 340,176 | 0 |
| Headloaders | 88,435 | 12 | 0 | 0 | 0 | 0 | 0 | 3,802 | 20,618 | 20,618 | 20,618 | 20,618 | 20,618 | 20,618 | 20,618 | 20,618 | 20,618 |
| Graziers | 435,859 | 12 | 62,423 | 76,260 | 76,260 | 71,955 | 56,888 | 52,583 | 52,583 | 52,583 | 46,125 | 41,820 | 41,820 | 41,820 | 41,820 | 41,820 | 41,820 |
| Collectors | 1,051,068 | 12 | 0 | 0 | 0 | 0 | 13,875 | 20,125 | 20,125 | 20,125 | 145,000 | 145,000 | 201,250 | 201,250 | 201,250 | 340,000 | 402,500 |
| Total | 2,086,840 | 12 | 62,423 | 123,258 | 136,686 | 71,955 | 74,120 | 264,501 | 93,326 | 138,708 | 218,457 | 502,854 | 331,762 | 263,688 | 272,640 | 742,614 | 464,938 |
| Non-revenue | 1,575,362 | 12 | 62,423 | 76,260 | 76,260 | 71,955 | 70,763 | 76,509 | 93,326 | 93,326 | 211,743 | 207,438 | 263,688 | 263,688 | 263,688 | 402,438 | 464,938 |
| Economic | 1,996,988 | 12 | 60,648 | 149,631 | 135,087 | -155,145 | -109,524 | 60,093 | -51,574 | 138,691 | 215,571 | 790,270 | 399,836 | 263,688 | 281,592 | 1,082,790 | 464,938 |

**Table MT:10  (B) Years 16-34**  Forest Productivity Model (Value of Total Plantation Outputs)

| Plantation | Output | Unit | 16 | 17 | 18 | 19 | 20 | 21 | 22 | 23 | 24 | 25 | 26 | 27 | 28 | 29 | 30 | 31 | 32 | 33 | 34 |
|---|---|---|---|---|---|---|---|---|---|---|---|---|---|---|---|---|---|---|---|---|---|
| Teak | Timber | m^3 | 0 | 0 | 0 | 0 | 0 | 0 | 0 | 0 | 0 | 0 | 0 | 0 | 0 | 0 | 0 | 0 | 0 | 0 | 0 |
| | Medium poles | pole | 0 | 0 | 0 | 0 | 0 | 0 | 0 | 0 | 0 | 0 | 0 | 0 | 0 | 0 | 0 | 0 | 0 | 0 | 0 |
| | Small poles | pole | 0 | 0 | 0 | 0 | 0 | 0 | 0 | 0 | 0 | 45,304 | 0 | 0 | 0 | 0 | 0 | 0 | 0 | 0 | 0 |
| | Fuelwood | m^3 | 0 | 0 | 0 | 0 | 0 | 0 | 30,203 | 0 | 0 | 0 | 0 | 0 | 0 | 0 | 0 | 0 | 0 | 0 | 0 |
| Sewan | Timber | m^3 | 0 | 0 | 0 | 0 | 0 | 0 | 0 | 0 | 0 | 0 | 0 | 0 | 0 | 0 | 0 | 0 | 0 | 0 | 0 |
| | Fuelwood | m^3 | 0 | 0 | 0 | 0 | 0 | 0 | 0 | 0 | 0 | 0 | 0 | 0 | 0 | 0 | 0 | 0 | 0 | 0 | 0 |
| Non-teak Fuelwood | | m^3 | 20,618 | 20,618 | 20,618 | 20,618 | 20,618 | 20,618 | 20,618 | 20,618 | 20,618 | 20,618 | 20,618 | 20,618 | 20,618 | 20,618 | 20,618 | 20,618 | 20,618 | 20,618 | 20,618 |
| Bamboo Poles | Poles | Poles | 0 | 16,000 | 608,000 | 0 | 0 | 16,000 | 608,000 | 0 | 0 | 16,000 | 608,000 | 0 | 0 | 16,000 | 608,000 | 0 | 0 | 16,000 | 608,000 |
| | Lops and tops | Kg | 0 | 1,904 | 72,352 | 0 | 0 | 1,904 | 72,352 | 0 | 0 | 1,904 | 72,352 | 0 | 0 | 1,904 | 72,352 | 0 | 0 | 1,904 | 72,352 |
| MFP | Fruit/Medicinal plants | Kg | 402,500 | 402,500 | 402,500 | 541,250 | 603,750 | 603,750 | 603,750 | 603,750 | 742,500 | 805,000 | 805,000 | 250,000 | 0 | 0 | 0 | 0 | 0 | 0 | 0 |
| | Fuelwood | m^3 | 0 | 0 | 0 | 0 | 0 | 0 | 0 | 0 | 0 | 0 | 733 | 330 | 0 | 0 | 0 | 0 | 0 | 0 | 0 |
| Stylo | | | 24,600 | 24,600 | 24,600 | 24,600 | 24,600 | 24,600 | 24,600 | 24,600 | 24,600 | 24,600 | 24,600 | 24,600 | 24,600 | 24,600 | 24,600 | 24,600 | 24,600 | 24,600 | 24,600 |
| Grazing | | MT | 0 | 0 | 0 | 0 | 0 | 0 | 0 | 0 | 0 | 0 | 0 | 0 | 0 | 0 | 0 | 0 | 0 | 0 | 0 |
| **Total** | | | 447,718 | 465,622 | 1,128,070 | 586,468 | 648,968 | 666,872 | 1,359,523 | 648,968 | 787,718 | 913,426 | 1,531,303 | 295,548 | 45,218 | 63,122 | 725,570 | 45,218 | 45,218 | 63,122 | 725,570 |

**Distribution of Benefit Flows From Plantations Under JFM**

| | | | 16 | 17 | 18 | 19 | 20 | 21 | 22 | 23 | 24 | 25 | 26 | 27 | 28 | 29 | 30 | 31 | 32 | 33 | 34 |
|---|---|---|---|---|---|---|---|---|---|---|---|---|---|---|---|---|---|---|---|---|---|
| FD | Revenue | | 0 | 8,952 | 340,176 | 0 | 0 | 8,952 | 355,277 | 0 | 0 | 31,604 | 340,176 | 0 | 0 | 8,952 | 340,176 | 0 | 0 | 8,952 | 340,176 |
| | - Costs | | 0 | 0 | 0 | 0 | 0 | 0 | 0 | 0 | 0 | 0 | 0 | 0 | 0 | 0 | 0 | 0 | 0 | 0 | 0 |
| | Net revenue | | 0 | 8,952 | 340,176 | 0 | 0 | 8,952 | 355,277 | 0 | 0 | 31,604 | 340,176 | 0 | 0 | 8,952 | 340,176 | 0 | 0 | 8,952 | 340,176 |
| Vill | Revenue | | 0 | 8,952 | 340,176 | 0 | 0 | 8,952 | 355,277 | 0 | 0 | 31,604 | 340,176 | 0 | 0 | 8,952 | 340,176 | 0 | 0 | 8,952 | 340,176 |
| | Headloaders | | 20,618 | 20,618 | 20,618 | 20,618 | 20,618 | 20,618 | 20,618 | 20,618 | 20,618 | 20,618 | 21,351 | 20,948 | 20,618 | 20,618 | 20,618 | 20,618 | 20,618 | 20,618 | 20,618 |
| | Graziers | | 24,600 | 24,600 | 24,600 | 24,600 | 24,600 | 24,600 | 24,600 | 24,600 | 24,600 | 24,600 | 24,600 | 24,600 | 24,600 | 24,600 | 24,600 | 24,600 | 24,600 | 24,600 | 24,600 |
| | Collectors | | 402,500 | 402,500 | 402,500 | 541,250 | 603,750 | 603,750 | 603,750 | 603,750 | 742,500 | 805,000 | 805,000 | 250,000 | 0 | 0 | 0 | 0 | 0 | 0 | 0 |
| | Total | | 447,718 | 456,670 | 787,894 | 586,468 | 648,968 | 657,920 | 1,004,246 | 648,968 | 787,718 | 881,822 | 1,191,127 | 295,548 | 45,218 | 54,170 | 385,394 | 45,218 | 45,218 | 54,170 | 385,394 |
| | Non-revenue | | 447,718 | 447,718 | 447,718 | 586,468 | 648,968 | 648,968 | 648,968 | 648,968 | 787,718 | 850,218 | 850,951 | 295,548 | 45,218 | 45,218 | 45,218 | 45,218 | 45,218 | 45,218 | 45,218 |
| Economic | | | 447,718 | 465,622 | 1,128,070 | 586,468 | 648,968 | 666,872 | 1,359,523 | 648,968 | 787,718 | 913,426 | 1,531,303 | 295,548 | 45,218 | 63,122 | 725,570 | 45,218 | 45,218 | 63,122 | 725,570 |

**Table MT.10 (C) Years 34-50**  Forest Productivity Model (Value of Total Plantation Outputs)

| Plantation | Output | Unit | 35 | 36 | 37 | 38 | 39 | 40 | 41 | 42 | 43 | 44 | 45 | 46 | 47 | 48 | 49 | 50 |
|---|---|---|---|---|---|---|---|---|---|---|---|---|---|---|---|---|---|---|
| | | | | | | | | | | | | | | | | | | Year |
| Teak | Timber | m^3 | 0 | 0 | 0 | 0 | 0 | 0 | 0 | 0 | 0 | 0 | 0 | 0 | 299,664 | 0 | 0 | 249,720 |
| | Medium poles | pole | 0 | 0 | 0 | 0 | 0 | 0 | 0 | 0 | 0 | 0 | 0 | 0 | 0 | 0 | 0 | 0 |
| | Small poles | pole | 0 | 0 | 0 | 0 | 0 | 0 | 0 | 0 | 0 | 0 | 0 | 0 | 0 | 0 | 0 | 0 |
| | Fuelwood | m^3 | 0 | 0 | 0 | 0 | 0 | 0 | 0 | 0 | 0 | 0 | 0 | 0 | 10,368 | 0 | 0 | 8,640 |
| Sewan | Timber | m^3 | 0 | 0 | 0 | 0 | 0 | 0 | 0 | 0 | 0 | 0 | 0 | 53,550 | 0 | 0 | 0 | 0 |
| | Fuelwood | m^3 | 0 | 0 | 0 | 0 | 0 | 0 | 0 | 0 | 0 | 0 | 0 | 606 | 0 | 0 | 0 | 0 |
| Non-teak | Fuelwood | m^3 | 20,618 | 20,618 | 20,618 | 20,618 | 20,618 | 20,618 | 20,618 | 20,618 | 20,618 | 20,618 | 20,618 | 20,618 | 20,618 | 20,618 | 20,618 | 20,618 |
| Bamboo | Poles | Poles | 0 | 0 | 16,000 | 608,000 | 0 | 0 | 16,000 | 608,000 | 0 | 0 | 16,000 | 608,000 | 0 | 0 | 16,000 | 608,000 |
| | Lops and tops | Kg | 0 | 0 | 1,904 | 72,352 | 0 | 0 | 1,904 | 72,352 | 0 | 0 | 1,904 | 72,352 | 0 | 0 | 1,904 | 72,352 |
| MFP | Fruit/Medicinal plants | | 0 | 0 | 13,875 | 13,875 | 20,125 | 20,125 | 145,000 | 145,000 | 145,000 | 201,250 | 201,250 | 340,000 | 340,000 | 402,500 | 402,500 | 402,500 |
| | Fuelwood | m^3 | 0 | 0 | 0 | 0 | 0 | 0 | 0 | 0 | 0 | 0 | 0 | 0 | 0 | 0 | 0 | 0 |
| Stylo | Grazing | MT | 24,600 | 24,600 | 24,600 | 24,600 | 24,600 | 24,600 | 24,600 | 24,600 | 24,600 | 24,600 | 24,600 | 24,600 | 24,600 | 24,600 | 24,600 | 24,600 |
| | | | 0 | 0 | 0 | 0 | 0 | 0 | 0 | 0 | 0 | 0 | 0 | 0 | 0 | 0 | 0 | 0 |
| Total | | | 45,218 | 45,218 | 76,997 | 739,445 | 65,343 | 65,343 | 208,122 | 870,570 | 190,218 | 246,468 | 264,372 | 1,119,726 | 695,250 | 447,718 | 465,622 | 1,386,430 |

**Distribution of Benefit Flows From Plantations Under JFM**

| | | | 35 | 36 | 37 | 38 | 39 | 40 | 41 | 42 | 43 | 44 | 45 | 46 | 47 | 48 | 49 | 50 |
|---|---|---|---|---|---|---|---|---|---|---|---|---|---|---|---|---|---|---|
| FD | Revenue | | 0 | 0 | 8,952 | 340,176 | 0 | 0 | 8,952 | 340,176 | 0 | 0 | 8,952 | 367,254 | 155,016 | 0 | 8,952 | 469,356 |
| | - Costs | | 0 | 0 | 0 | 0 | 0 | 0 | 0 | 0 | 0 | 0 | 0 | 0 | 0 | 0 | 0 | 0 |
| | Net revenue | | 0 | 0 | 8,952 | 340,176 | 0 | 0 | 8,952 | 340,176 | 0 | 0 | 8,952 | 367,254 | 155,016 | 0 | 8,952 | 469,356 |
| Vill | Revenue | | 0 | 0 | 8,952 | 340,176 | 0 | 0 | 8,952 | 340,176 | 0 | 0 | 8,952 | 367,254 | 155,016 | 0 | 8,952 | 469,356 |
| | Headloaders | | 20,618 | 20,618 | 20,618 | 20,618 | 20,618 | 20,618 | 20,618 | 20,618 | 20,618 | 20,618 | 20,618 | 20,618 | 20,618 | 20,618 | 20,618 | 20,618 |
| | Graziers | | 24,600 | 24,600 | 24,600 | 24,600 | 24,600 | 24,600 | 24,600 | 24,600 | 24,600 | 24,600 | 24,600 | 24,600 | 24,600 | 24,600 | 24,600 | 24,600 |
| | Collectors | | 0 | 0 | 13,875 | 13,875 | 20,125 | 20,125 | 145,000 | 145,000 | 145,000 | 201,250 | 201,250 | 340,000 | 340,000 | 402,500 | 402,500 | 402,500 |
| | Total | | 45,218 | 45,218 | 68,045 | 399,269 | 65,343 | 65,343 | 199,170 | 530,394 | 190,218 | 246,468 | 255,420 | 752,472 | 540,234 | 447,718 | 456,670 | 917,074 |
| | Non-revenue | | 45,218 | 45,218 | 59,093 | 59,093 | 65,343 | 65,343 | 190,218 | 190,218 | 190,218 | 246,468 | 246,468 | 385,218 | 385,218 | 447,718 | 447,718 | 447,718 |
| Economic | | | 45,218 | 45,218 | 76,997 | 739,445 | 65,343 | 65,343 | 208,122 | 870,570 | 190,218 | 246,468 | 264,372 | 1,119,726 | 695,250 | 447,718 | 465,622 | 1,386,430 |

**Table MT.11  Forest Productivity Model (Outputs on Plantation Areas and Without JFM)**

| Output | Unit | Unit Prices (Rs) | 1 | 2 | 3 | 4 | 5 | 6 | 7 | 8 | 9 | 10 | 11 | 12-50 |
|---|---|---|---|---|---|---|---|---|---|---|---|---|---|---|
| **Natural Forest (NF-3)** | | | | | | | | | | | | | | |
| GS | m^3/ha | | 13.98 | 12.24 | 11.14 | 9.95 | 8.73 | 7.49 | 6.22 | 4.92 | 3.59 | 2.23 | 0.84 | 0.00 |
| CAI | m^3/ha | | 4.14 | 4.29 | 4.38 | 4.48 | 4.58 | 4.69 | 4.80 | 4.90 | 5.02 | 5.13 | 0.00 | 0.00 |
| CC | (%) | | 9 | 8 | 7 | 6 | 5 | 4 | 3 | 3 | 2 | 1 | 1 | 0 |
| Dead/dry | m^3/ha | 135.00 | 0.45 | 0.41 | 0.43 | 0.44 | 0.45 | 0.46 | 0.47 | 0.48 | 0.49 | 0.50 | 0.51 | 0.00 |
| Green cut | m^3/ha | 135.00 | 5.43 | 4.97 | 5.15 | 5.26 | 5.38 | 5.50 | 5.63 | 5.75 | 5.89 | 6.02 | 6.16 | 0.00 |
| Teak | m^3/ha | 8,324.00 | | | | | | | | | | | | |
| Fuelwood | m^3/ha | 1,152.00 | | | | | | | | | | | | |
| Non teak | m^3/ha | 459.00 | | | | | | | | | | | | |
| Fuelwood | m^3/ha | 135.00 | | | | | | | | | | | | |
| Bamboo Poles | Poles | 2.00 | 800 | 800 | 800 | 800 | 800 | 800 | 800 | 800 | 800 | 800 | 800 | 800 |
| Lops/tops | kg/pole | 0.11 | 3,400 | 3,400 | 3,400 | 3,400 | 3,400 | 3,400 | 3,400 | 3,400 | 3,400 | 3,400 | 3,400 | 3,400 |
| Grazing | MT/ha | 410.00 | 1.22 | 1.25 | 1.26 | 1.28 | 1.30 | 1.32 | 1.33 | 1.35 | 1.36 | 1.38 | 1.39 | 1.40 |
| Tendu | LS/ha | 1 | 120.00 | 120.00 | 120.00 | 120.00 | 120.00 | 120.00 | 120.00 | 120.00 | 120.00 | 120.00 | 120.00 | 120.00 |
| NTFP | | 1 | 0.00 | 0.00 | 0.00 | 0.00 | 0.00 | 0.00 | 0.00 | 0.00 | 0.00 | 0.00 | 0.00 | 0.00 |

**Total Flow of Benefits Without Plantation (102.5 ha)**

| Output | Unit | Unit Prices (Rs) | 1 | 2 | 3 | 4 | 5 | 6 | 7 | 8 | 9 | 10 | 11 | 12-50 |
|---|---|---|---|---|---|---|---|---|---|---|---|---|---|---|
| Teak | m^3 | | 0 | 0 | 0 | 0 | 0 | 0 | 0 | 0 | 0 | 0 | 0 | 0 |
| Fuelwood | m^3 | | 0 | 0 | 0 | 0 | 0 | 0 | 0 | 0 | 0 | 0 | 0 | 0 |
| Non teak | m^3 | | 0 | 0 | 0 | 0 | 0 | 0 | 0 | 0 | 0 | 0 | 0 | 0 |
| Fuelwood | m^3 | | 0 | 0 | 0 | 0 | 0 | 0 | 0 | 0 | 0 | 0 | 0 | 0 |
| Dead/dry | m^3/ha | 37,304 | 6,259 | 5,731 | 5,934 | 6,062 | 6,201 | 6,343 | 6,488 | 6,636 | 6,787 | 6,942 | 7,101 | 0 |
| Green cut | m^3 | 447,646 | 75,105 | 68,774 | 71,207 | 72,741 | 74,413 | 76,111 | 77,850 | 79,628 | 81,447 | 83,308 | 85,211 | 0 |
| Bamboo Poles | Poles | 1,351,979 | 164,000 | 164,000 | 164,000 | 164,000 | 164,000 | 164,000 | 164,000 | 164,000 | 164,000 | 164,000 | 164,000 | 164,000 |
| Lops/tops | | 316,025 | 38,335 | 38,335 | 38,335 | 38,335 | 38,335 | 38,335 | 38,335 | 38,335 | 38,335 | 38,335 | 38,335 | 38,335 |
| Grazing | MT | 458,074 | 51,209 | 52,409 | 53,126 | 53,873 | 54,594 | 55,293 | 55,968 | 56,615 | 57,233 | 57,818 | 58,367 | 58,677 |
| Tendu | LS | 101,398 | 12,300 | 12,300 | 12,300 | 12,300 | 12,300 | 12,300 | 12,300 | 12,300 | 12,300 | 12,300 | 12,300 | 12,300 |
| NTFP | LS | 0 | 0 | 0 | 0 | 0 | 0 | 0 | 0 | 0 | 0 | 0 | 0 | 0 |
| Total | | 2,712,427 | 347,208 | 341,549 | 344,902 | 347,310 | 349,843 | 352,382 | 354,940 | 357,514 | 360,102 | 362,703 | 365,314 | 273,312 |

**Distribution of Total Flow of Benefits Without JFM**

| | | Net Worth (Rs) | Discount Rate (%) | 1 | 2 | 3 | 4 | 5 | 6 | 7 | 8 | 9 | 10 | 11 | 12-50 |
|---|---|---|---|---|---|---|---|---|---|---|---|---|---|---|---|---|
| FD | Revenue | 1,680,291 | * | 202,335 | 202,335 | 202,335 | 202,335 | 202,335 | 202,335 | 202,335 | 202,335 | 202,335 | 202,335 | 202,335 | 202,335 |
| Vil | Revenue | 0 | * | 0 | 0 | 0 | 0 | 0 | 0 | 0 | 0 | 0 | 0 | 0 | 0 |
| | Headloaders | 484,950 | * | 81,364 | 74,505 | 77,141 | 78,803 | 80,614 | 82,454 | 84,338 | 86,264 | 88,235 | 90,250 | 92,312 | 0 |
| | Graziers | 461,637 | * | 51,209 | 52,409 | 53,126 | 53,873 | 54,594 | 55,293 | 55,968 | 56,615 | 57,233 | 57,818 | 58,367 | 58,677 |
| | Collectors | 102,145 | * | 12,300 | 12,300 | 12,300 | 12,300 | 12,300 | 12,300 | 12,300 | 12,300 | 12,300 | 12,300 | 12,300 | 12,300 |
| | Total village | 1,048,733 | * | 144,873 | 139,214 | 142,567 | 144,975 | 147,508 | 150,047 | 152,605 | 155,179 | 157,767 | 160,368 | 162,979 | 70,977 |
| | Non-revenue | 1,048,733 | * | 144,873 | 139,214 | 142,567 | 144,975 | 147,508 | 150,047 | 152,605 | 155,179 | 157,767 | 160,368 | 162,979 | 70,977 |
| Economic | | 2,729,023 | * | 347,208 | 341,549 | 344,902 | 347,310 | 349,843 | 352,382 | 354,940 | 357,514 | 360,102 | 362,703 | 365,314 | 273,312 |

**Table MT.12 — Incremental benefits to Villages from Plantations under JFM**

| | Net Worth (Rs, 50 yrs) | Discount Rate (%) | Year 1 | 2 | 3 | 4 | 5 | 6 | 7 | 8 | 9 | 10 | 11 | 12 | 13 | 14 | 15 |
|---|---|---|---|---|---|---|---|---|---|---|---|---|---|---|---|---|---|
| **FD** | | | | | | | | | | | | | | | | | |
| Revenue | -1,770,143 | 12 | -204,110 | -175,962 | -203,934 | -429,435 | -385,978 | -406,743 | -347,235 | -202,353 | -205,221 | 85,081 | -134,261 | -202,335 | -193,383 | 137,841 | -202,335 |
| **Village** | | | | | | | | | | | | | | | | | |
| Revenue | 511,479 | 12 | 0 | 46,998 | 60,426 | 0 | 3,357 | 187,992 | 0 | 45,383 | 6,714 | 295,416 | 68,074 | 0 | 8,952 | 340,176 | 0 |
| Headloaders | -396,515 | 12 | -81,364 | -74,505 | -77,141 | -78,803 | -80,614 | -78,652 | -63,719 | -65,646 | -67,616 | -69,632 | -71,693 | 20,618 | 20,618 | 20,618 | 20,618 |
| Graziers | -25,778 | 12 | 11,213 | 23,851 | 23,134 | 18,082 | 2,294 | -2,711 | -3,385 | -4,032 | -11,108 | -15,998 | -16,547 | -16,857 | -16,857 | -16,857 | -16,857 |
| Collectors | 948,922 | 12 | -12,300 | -12,300 | -12,300 | -12,300 | 1,575 | 7,825 | 7,825 | 7,825 | 132,700 | 132,700 | 188,950 | 188,950 | 188,950 | 327,700 | 390,200 |
| Total village | 1,038,108 | 12 | -82,451 | -15,956 | -5,881 | -73,020 | -73,388 | 114,454 | -59,279 | -16,471 | 60,690 | 342,487 | 168,783 | 192,712 | 201,664 | 671,638 | 393,962 |
| Non-revenue | 526,629 | 12 | -82,451 | -62,954 | -66,307 | -73,020 | -76,745 | -73,538 | -59,279 | -61,853 | 53,976 | 47,071 | 100,710 | 192,712 | 192,712 | 331,462 | 393,962 |
| **Economic** | | | | | | | | | | | | | | | | | |
| Net Worth | -732,036 | 12 | -286,561 | -191,918 | -209,815 | -502,455 | -459,366 | -292,289 | -406,514 | -218,823 | -144,531 | 427,568 | 34,522 | -9,623 | 8,281 | 809,479 | 191,627 |
| IRR (%) | | 7 % | | | | | | | | | | | | | | | |

| | Year 16 | 17 | 18 | 19 | 20 | 21 | 22 | 23 | 24 | 25 | 26 | 27 | 28 | 29 | 30 | 31 | 32 |
|---|---|---|---|---|---|---|---|---|---|---|---|---|---|---|---|---|---|
| **FD** | | | | | | | | | | | | | | | | | |
| Revenue | -202,335 | -193,383 | 137,841 | -202,335 | -202,335 | -193,383 | 152,942 | -202,335 | -202,335 | -170,731 | 137,841 | -202,335 | -202,335 | -193,383 | 137,841 | -202,335 | -202,335 |
| **Village** | | | | | | | | | | | | | | | | | |
| Revenue | 0 | 8,952 | 340,176 | 0 | 0 | 8,952 | 355,277 | 0 | 0 | 31,604 | 340,176 | 0 | 0 | 8,952 | 340,176 | 0 | 0 |
| Headloaders | 20,618 | 20,618 | 20,618 | 20,618 | 20,618 | 20,618 | 20,618 | 20,618 | 20,618 | 20,618 | 21,351 | 20,948 | 20,618 | 20,618 | 20,618 | 20,618 | 20,618 |
| Graziers | -34,077 | -34,077 | -34,077 | -34,077 | -34,077 | -34,077 | -34,077 | -34,077 | -34,077 | -34,077 | -34,077 | -34,077 | -34,077 | -34,077 | -34,077 | -34,077 | -34,077 |
| Collectors | 390,200 | 390,200 | 390,200 | 528,950 | 591,450 | 591,450 | 591,450 | 591,450 | 730,200 | 792,700 | 792,700 | 237,700 | -12,300 | -12,300 | -12,300 | -12,300 | -12,300 |
| Total village | 376,742 | 385,694 | 716,918 | 515,492 | 577,992 | 586,944 | 933,269 | 577,992 | 716,742 | 810,845 | 1,120,150 | 224,572 | -25,758 | -16,806 | 314,418 | -25,758 | -25,758 |
| Non-revenue | 376,742 | 376,742 | 376,742 | 515,492 | 577,992 | 577,992 | 577,992 | 577,992 | 716,742 | 779,242 | 779,974 | 224,572 | -25,758 | -25,758 | 314,418 | -25,758 | -25,758 |
| **Economic** | 174,407 | 192,311 | 854,759 | 313,157 | 375,657 | 393,561 | 1,086,211 | 375,657 | 514,407 | 640,114 | 1,257,991 | 22,237 | -228,093 | -210,189 | 452,259 | -228,093 | -228,093 |

| | Year 33 | 34 | 35 | 36 | 37 | 38 | 39 | 40 | 41 | 42 | 43 | 44 | 45 | 46 | 47 | 48 | 49 | 50 |
|---|---|---|---|---|---|---|---|---|---|---|---|---|---|---|---|---|---|---|
| **FD** | | | | | | | | | | | | | | | | | | |
| Revenue | -193,383 | 137,841 | -202,335 | -202,335 | -193,383 | 137,841 | -202,335 | -202,335 | -193,383 | 137,841 | -202,335 | -202,335 | -193,383 | 164,919 | -47,319 | -202,335 | -193,383 | 267,021 |
| **Village** | | | | | | | | | | | | | | | | | | |
| Revenue | 8,952 | 340,176 | 0 | 0 | 8,952 | 340,176 | 0 | 0 | 8,952 | 340,176 | 0 | 0 | 8,952 | 367,254 | 155,016 | 0 | 8,952 | 469,356 |
| Headloaders | 20,618 | 20,618 | 20,618 | 20,618 | 20,618 | 20,618 | 20,618 | 20,618 | 20,618 | 20,618 | 20,618 | 20,618 | 20,618 | 20,618 | 20,618 | 20,618 | 20,618 | 20,618 |
| Graziers | -34,077 | -34,077 | -34,077 | -34,077 | -34,077 | -34,077 | -34,077 | -34,077 | -34,077 | -34,077 | -34,077 | -34,077 | -34,077 | -34,077 | -34,077 | -34,077 | -34,077 | -34,077 |
| Collectors | -12,300 | -12,300 | -12,300 | -12,300 | 1,575 | 1,575 | 7,825 | 7,825 | 132,700 | 132,700 | 132,700 | 188,950 | 188,950 | 327,700 | 327,700 | 390,200 | 390,200 | 390,200 |
| Total village | -16,806 | 314,418 | 385,694 | -25,758 | -2,931 | 328,293 | -5,633 | -5,633 | 128,194 | 459,418 | 119,242 | 175,492 | 184,444 | 681,496 | 469,258 | 376,742 | 385,694 | 846,098 |
| Non-revenue | -25,758 | -25,758 | -25,758 | -25,758 | -11,883 | -11,883 | -5,633 | -5,633 | 119,242 | 119,242 | 119,242 | 175,492 | 175,492 | 314,242 | 314,242 | 376,742 | 376,742 | 376,742 |
| **Economic** | -210,189 | 452,259 | 192,311 | -228,093 | -196,314 | 466,134 | -207,968 | -207,968 | -65,189 | 597,259 | -83,093 | -26,843 | -8,939 | 846,414 | 421,939 | 174,407 | 192,311 | 1,113,119 |

**Table MT.13**          **Teak productivity**

| Year | GS Final | GS Total | CAI Total | CIP (%) | 1/GS Final |
|---|---|---|---|---|---|
| 1 | 5 | 5 | 5.000 | 100.00 | 0.200 |
| 5 | 15 | 15 | 3.000 | 20.00 | 0.067 |
| 10 | 26.94 | 26.94 | 2.694 | 10.00 | 0.037 |
| 15 | 34.99 | 42.33 | 3.078 | 7.27 | 0.029 |
| 20 | 41.28 | 55.63 | 2.660 | 4.78 | 0.024 |
| 25 | 48.48 | 66.12 | 2.098 | 3.17 | 0.021 |
| 30 | 50.38 | 76.62 | 2.100 | 2.74 | 0.020 |
| 35 | 53.18 | 84.32 | 1.540 | 1.83 | 0.019 |
| 40 | 58.08 | 92 | 1.536 | 1.67 | 0.017 |
| 45 | 61.23 | 100.06 | 1.612 | 1.61 | 0.016 |
| 50 | 65.78 | 107.76 | 1.540 | 1.43 | 0.015 |

Source:    Standard yield tables (Site-quality IV)

        GS (Final) =    Standing Growing Stock
        GS (Total) =    Standing Growing Stock + Intermediate Yields

Relationships

$$CIP\,(\%) = -8.43 + 532 \times (1/GS)$$

**Figure MT.1  CIP (%) versus Growing Stock**

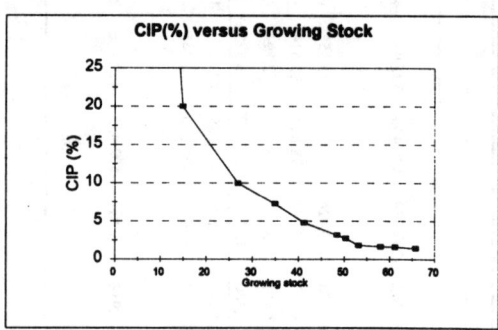

**Regression Output:**

| | |
|---|---|
| Constant | -8.43 |
| Std Err of Y Est | 2.63 |
| R Squared | 0.99 |
| No. of Observations | 11 |
| Degrees of Freedom | 9 |
| | |
| X Coefficient(s) | 532.10 |
| Std Err of Coef. | 15.31 |

92

**Table MT.14**      **Estimated Offtake Rates**

| Forest Type | Area (ha) | Degradation Status | Protection | Estimated Current Offtake %CAI |
|---|---|---|---|---|
| NF.1 | 30 | Moderate | With JFM | 10 |
|  |  |  | Without JFM | 30 |
| NF.2 | 20 | Moderate | With JFM | 10 |
|  |  |  | Without JFM | 50 |
| NF.3 | 20 | Partial | With JFM | 10 |
|  |  |  | Without JFM | 90 |
| NF.4 | 20 | High | With JFM | 10 |
|  |  |  | Without JFM | 120 |

Note:    Offtake without JFM may amount to more than 100% of CAI, as it includes intermediate products

**Table MT.15**          **Projected Yields in Natural Forest**

| Sample Area | Management | Measured Standing Stock | | | Offtake | Projected Standing Stock | | Projected Total Yield | |
|---|---|---|---|---|---|---|---|---|---|
| | | Age | GS | MAI | CIP | GS | MAI | GS | MAI |
| NF.1 | With JFM | 25 | 32.5 | 1.3 | 10 | 58.8 | 1.78 | 58.8 | 1.78 |
| | Without JFM | | | | 30 | 56.4 | 1.28 | 63.1 | 1.26 |
| NF.2 | With JFM | 25 | 21.7 | 0.87 | 10 | 57.3 | 1.15 | 57.3 | 1.15 |
| | Without JFM | | | | 50 | 49 | 0.98 | 70.3 | 1.41 |
| NF.3 | With JFM | 15 | 27.6 | 1.84 | 10 | 60.8 | 1.22 | 60.8 | 1.22 |
| | Without JFM | | | | 90 | 38.1 | 0.76 | 106.6 | 2.13 |
| NF.4 | With JFM | 15 | 3.7 | 0.25 | 10 | 60.3 | 1.21 | 60.3 | 1.21 |
| | Without JFM | | | | 120 | 0 | 0 | 129.7 | 2.59 |

**Table MT.16**     **Relationship between Canopy Cover (CC) and Growing stock (GS)**

| Canopy Cover CC | Growing Stock GS | GS^2 |
|---|---|---|
| 0 | 0.0 | 0 |
| 20 | 24.0 | 576 |
| 40 | 40.0 | 1,600 |
| 60 | 50.0 | 2,500 |

Relationships

$$CC = 0.188 + 0.424\ GS + 0.015\ GS^2$$

Regression Output:

| | | |
|---|---|---|
| Constant | 0.19 | |
| Std Err of Y Est | 1.82 | |
| R Squared | 1.00 | |
| No. of Observations | 4 | |
| Degrees of Freedom | 1 | |
| | | |
| X Coefficient(s) | 0.42359 | 0.01518 |
| Std Err of Coef. | 0.17349 | 0.00342 |

**Figure MT.2  Canopy Cover versus Growing Stock**

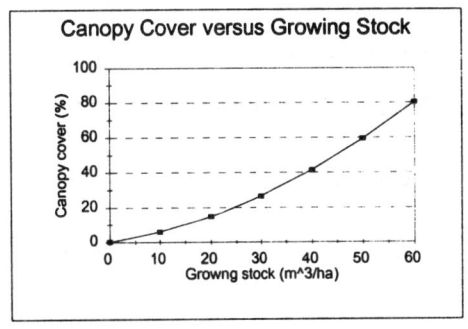

**Table MT.17    Grass Productivity in Natural Forest**

| Basal Area (m^2/ha) | Canopy Cover (%) | Grass (MT) |
|---|---|---|
| 0 | 0 | 2.2 (1) |
| 0 | 0 | 1.4 |
| 4 | 0.1 | 1.2 |
| 6 | 0.2 | 1.0 |
| 9 | 0.3 | 0.8 |
| 12 | 0.4 | 0.6 |
| 16 | 0.5 | 0.4 |
| 20 | 0.6 | 0.4 |
| 25 | 0.8 | 0.4 |

Note (1)  Without rootstock

Relationships

Grass = 1.5 -(10/6) x  Basal area

**Figure MT.3 Basal Area versus Canopy Cover**

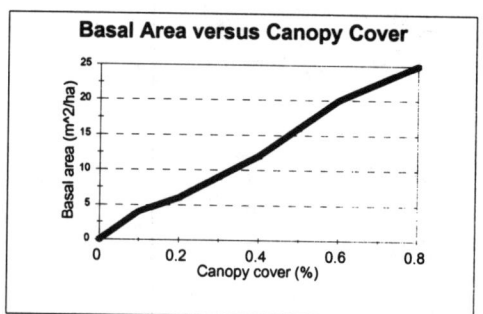

Grass = 1.4 - .02 x Canopy cover

**Figure MT.4 Grass production versus Canopy Cov**

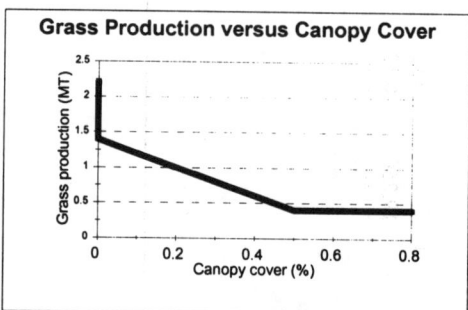

**Table MT.18**   Incremental Benefits to Village from Natural Forest (90ha)

| Management | | Net Worth (Rs) | Discount Rate (%) | Year | | | | | | | | | | |
|---|---|---|---|---|---|---|---|---|---|---|---|---|---|---|
| | | | | 1 | 2 | 6 | 10 | 14 | 18 | 22 | 25 | 30 | 34 | 35 |
| **With JFM** | | | | | | | | | | | | | | |
| FD | Revenue | 1,730,849 | 0.12 | 0 | 751,968 | 751,968 | 751,968 | 751,968 | 751,968 | 751,968 | 979,063 | 751,968 | 751,968 | 2,976,171 |
| VFC | Revenue | 1,730,849 | 0.12 | 0 | 751,968 | 751,968 | 751,968 | 751,968 | 751,968 | 751,968 | 979,063 | 751,968 | 751,968 | 2,976,171 |
| | Headloader | 19,250 | 0.12 | 3,804 | 3,483 | 2,548 | 1,864 | 1,364 | 998 | 730 | 577 | 186 | 136 | 126 |
| | Graziers | 175,384 | 0.12 | 36,624 | 33,824 | 23,371 | 14,564 | 7,502 | 2,003 | -2,198 | -4,657 | 28,891 | 28,053 | 27,880 |
| | Collecters | 218,689 | 0.12 | 14,000 | 15,700 | 22,900 | 31,100 | 40,100 | 44,100 | 44,100 | 44,100 | 25,600 | 23,600 | 23,600 |
| Total | Revenue | 2,144,172 | 0.12 | 54,428 | 804,975 | 800,787 | 799,497 | 800,933 | 799,068 | 794,600 | 1,019,083 | 806,645 | 803,757 | 3,027,777 |
| | Non-reven | 413,323 | 0.12 | 54,428 | 53,007 | 48,819 | 47,529 | 48,965 | 47,100 | 42,632 | 40,020 | 54,677 | 51,789 | 51,606 |
| Economic | | 3,875,021 | 0.12 | 54,428 | 1,556,943 | 1,552,755 | 1,551,465 | 1,552,901 | 1,551,036 | 1,546,568 | 1,998,146 | 1,558,613 | 1,555,725 | 6,003,948 |
| **Without JFM** | | | | | | | | | | | | | | |
| FD | Revenue | 602,924 | 0.12 | 64,155 | 64,155 | 64,155 | 64,155 | 64,155 | 64,155 | 64,155 | 64,155 | 64,155 | 64,155 | 4,204,958 |
| VFC | Revenue | 0 | 0.12 | 0 | 0 | 0 | 0 | 0 | 0 | 0 | 0 | 0 | 0 | 0 |
| | Headloader | 97,954 | 0.12 | 15,134 | 13,858 | 12,439 | 11,202 | 10,156 | 9,269 | 8,510 | 8,013 | 7,296 | 6,807 | 6,694 |
| | Graziers | 146,312 | 0.12 | 24,042 | 23,152 | 19,502 | 16,225 | 13,339 | 10,830 | 8,667 | 7,248 | 5,220 | 3,859 | 3,550 |
| | Collecters | 138,686 | 0.12 | 11,500 | 11,500 | 13,900 | 20,100 | 21,700 | 24,700 | 26,700 | 26,700 | 28,700 | 28,700 | 28,700 |
| Total | Revenue | 382,952 | 0.12 | 50,675 | 48,510 | 45,840 | 47,527 | 45,195 | 44,799 | 43,877 | 41,961 | 41,216 | 39,366 | 38,945 |
| | Non-reven | 382,952 | 0.12 | 50,675 | 48,510 | 45,840 | 47,527 | 45,195 | 44,799 | 43,877 | 41,961 | 41,216 | 39,366 | 38,945 |
| Economic | | 985,876 | 0.12 | 114,830 | 112,665 | 109,995 | 111,682 | 109,350 | 108,954 | 108,032 | 106,116 | 105,371 | 103,521 | 4,243,903 |
| **Incremental Benefits due to JFM** | | | | | | | | | | | | | | |
| FD | Revenue | 1,127,925 | 0.12 | -64,155 | 687,813 | 687,813 | 687,813 | 687,813 | 687,813 | 687,813 | 914,908 | 687,813 | 687,813 | -1,228,787 |
| VFC | Revenue | 1,730,849 | 0.12 | 0 | 751,968 | 751,968 | 751,968 | 751,968 | 751,968 | 751,968 | 979,063 | 751,968 | 751,968 | 2,976,171 |
| | Headloader | -78,703 | 0.12 | -11,330 | -10,375 | -9,890 | -9,337 | -8,793 | -8,271 | -7,781 | -7,436 | -7,110 | -6,670 | -6,568 |
| | Graziers | 29,072 | 0.12 | 12,583 | 10,672 | 3,869 | -1,660 | -5,837 | -8,827 | -10,864 | -11,904 | 23,671 | 24,194 | 24,330 |
| | Collecters | 80,002 | 0.12 | 2,500 | 4,200 | 9,000 | 11,000 | 18,400 | 19,400 | 17,400 | 17,400 | -3,100 | -5,100 | -5,100 |
| Total | Revenue | 1,761,220 | 0.12 | 3,752 | 756,465 | 754,947 | 751,970 | 755,738 | 754,269 | 750,723 | 977,123 | 765,429 | 764,392 | 2,988,832 |
| | Non-reven | 30,371 | 0.12 | 3,752 | 4,497 | 2,979 | 2 | 3,770 | 2,301 | -1,245 | -1,940 | 13,461 | 12,424 | 12,661 |
| Economic | | 2,889,145 | 0.12 | -60,403 | 1,444,278 | 1,442,760 | 1,439,783 | 1,443,551 | 1,442,082 | 1,438,536 | 1,892,031 | 1,453,242 | 1,452,205 | 1,760,045 |

**Table MT.19**   (A) Years 1-15   Incremental Benefits to Village from Natural Forest and Plantation

| Management | | Net Worth (Rs) | Discount Rate (%) | Year 1 | 2 | 3 | 4 | 5 | 6 | 7 | 8 | 9 | 10 | 11 | 12 | 13 | 14 | 15 |
|---|---|---|---|---|---|---|---|---|---|---|---|---|---|---|---|---|---|---|
| **With JFM** | | | | | | | | | | | | | | | | | | |
| FD | Revenue | 1,640,996 | 12 | -1,775 | 778,341 | -1,599 | -227,100 | -183,643 | 547,560 | -144,900 | -18 | -2,886 | 1,039,384 | 68,074 | 0 | 8,952 | 1,092,144 | 0 |
| Vill | Revenue | 2,242,328 | 12 | 0 | 798,966 | 60,426 | 0 | 3,357 | 939,960 | 0 | 45,383 | 6,714 | 1,047,384 | 68,074 | 0 | 8,952 | 1,092,144 | 0 |
| | Headloaders | 107,685 | 12 | 3,804 | 3,483 | 3,222 | 2,979 | 2,755 | 6,350 | 22,975 | 22,798 | 22,634 | 22,483 | 22,342 | 22,213 | 22,093 | 21,982 | 21,880 |
| | Graziers | 611,244 | 12 | 99,047 | 110,084 | 107,341 | 100,372 | 82,732 | 75,953 | 73,586 | 71,328 | 62,725 | 56,384 | 54,460 | 52,643 | 50,931 | 49,322 | 47,810 |
| | Collectors | 1,269,756 | 12 | 14,000 | 15,700 | 18,100 | 19,700 | 36,775 | 43,025 | 47,625 | 49,225 | 176,100 | 176,100 | 234,350 | 239,350 | 241,350 | 380,100 | 444,600 |
| Total | | 4,231,013 | 12 | 116,850 | 928,233 | 189,089 | 123,052 | 125,619 | 1,065,288 | 144,186 | 188,734 | 268,173 | 1,302,351 | 379,226 | 314,206 | 323,326 | 1,543,548 | 514,290 |
| Non-revenu | | 1,988,685 | 12 | 116,850 | 129,267 | 128,663 | 123,052 | 122,262 | 125,328 | 144,186 | 143,351 | 261,459 | 254,967 | 311,152 | 314,206 | 314,374 | 451,404 | 514,290 |
| Economic | | 5,872,009 | 12 | 115,075 | 1,706,574 | 187,490 | -104,048 | -58,024 | 1,612,848 | -714 | 188,716 | 265,287 | 2,341,735 | 447,299 | 314,206 | 332,278 | 2,635,692 | 514,290 |
| **Without JFM** | | | | | | | | | | | | | | | | | | |
| FD | Revenue | 2,283,215 | 12 | 266,490 | 266,490 | 266,490 | 266,490 | 266,490 | 266,490 | 266,490 | 266,490 | 266,490 | 266,490 | 266,490 | 266,490 | 266,490 | 266,490 | 266,490 |
| Vil | Revenue | 0 | 12 | 0 | 0 | 0 | 0 | 0 | 0 | 0 | 0 | 0 | 0 | 0 | 0 | 0 | 0 | 0 |
| | Headloaders | 582,904 | 12 | 96,498 | 88,363 | 90,662 | 91,944 | 93,396 | 94,893 | 96,447 | 98,058 | 99,726 | 101,452 | 103,235 | 10,657 | 10,402 | 10,156 | 9,921 |
| | Graziers | 607,949 | 12 | 75,251 | 75,561 | 75,331 | 75,155 | 74,974 | 74,795 | 74,614 | 74,430 | 74,240 | 74,042 | 73,834 | 73,410 | 72,701 | 72,016 | 71,354 |
| | Collectors | 240,832 | 12 | 23,800 | 23,800 | 23,800 | 26,200 | 26,200 | 26,200 | 27,800 | 27,800 | 30,800 | 32,400 | 32,400 | 32,400 | 34,000 | 34,000 | 37,000 |
| Total village | | 1,431,685 | 12 | 195,549 | 187,724 | 189,792 | 193,299 | 194,571 | 195,887 | 198,861 | 200,287 | 204,766 | 207,894 | 209,470 | 116,468 | 117,103 | 116,172 | 118,275 |
| Non-revenu | | 1,431,685 | 12 | 195,549 | 187,724 | 189,792 | 193,299 | 194,571 | 195,887 | 198,861 | 200,287 | 204,766 | 207,894 | 209,470 | 116,468 | 117,103 | 116,172 | 118,275 |
| Economic | | 3,714,900 | 12 | 462,039 | 454,214 | 456,282 | 459,789 | 461,061 | 462,377 | 465,351 | 466,777 | 471,256 | 474,384 | 475,960 | 382,958 | 383,593 | 382,662 | 384,765 |
| **Incremental benefits due to JFM** | | | | | | | | | | | | | | | | | | |
| FD | Revenue | -642,219 | 12 | -268,265 | 511,851 | -268,089 | -493,590 | -450,133 | 281,070 | -411,390 | -266,508 | -269,376 | 772,894 | -198,416 | -266,490 | -257,538 | 825,654 | -266,490 |
| Vil | Revenue | 2,242,328 | 12 | 0 | 798,966 | 60,426 | 0 | 3,357 | 939,960 | 0 | 45,383 | 6,714 | 1,047,384 | 68,074 | 0 | 8,952 | 1,092,144 | 0 |
| | Headloaders | -475,218 | 12 | -92,694 | -84,880 | -87,440 | -88,965 | -90,641 | -88,543 | -73,472 | -75,260 | -77,092 | -78,969 | -80,893 | 11,555 | 11,691 | 11,826 | 11,959 |
| | Graziers | 3,294 | 12 | 23,796 | 34,523 | 32,011 | 25,218 | 7,757 | 1,159 | -1,028 | -3,101 | -11,515 | -17,658 | -19,375 | -20,767 | -21,770 | -22,694 | -23,544 |
| | Collectors | 1,028,925 | 12 | -9,800 | -8,100 | -5,700 | -6,500 | 10,575 | 16,825 | 19,825 | 21,425 | 145,300 | 143,700 | 201,950 | 206,950 | 207,350 | 346,100 | 407,600 |
| Total village | | 2,799,328 | 12 | -78,698 | 740,509 | -703 | -70,247 | -68,952 | 869,401 | -54,674 | -11,553 | 63,407 | 1,094,457 | 169,756 | 197,738 | 206,223 | 1,427,375 | 396,014 |
| Non-revenu | | 557,000 | 12 | -78,698 | -58,457 | -61,129 | -70,247 | -72,309 | -70,559 | -54,674 | -56,936 | 56,693 | 47,073 | 101,682 | 197,738 | 197,271 | 335,231 | 396,014 |
| Economic | | 2,157,109 | 12 | -346,963 | 1,252,360 | -268,792 | -563,837 | -519,085 | 1,150,471 | -466,064 | -278,061 | -205,969 | 1,867,351 | -28,660 | -68,752 | -51,315 | 2,253,029 | 129,524 |
| Employment | | | | 1,050 | 18,775 | 51,925 | 218,200 | 163,700 | 382,000 | 132,900 | 45,400 | 9,600 | 8,000 | 0 | 0 | 0 | 0 | 0 |

**Table MT.19    (B) Years 16-34    Incremental Benefits to Village from Natural Forest and Plantation**

| Management | | 16 | 17 | 18 | 19 | 20 | 21 | 22 | 23 | 24 | 25 | 26 | 27 | 28 | 29 | 30 | 31 | 32 | 33 | 34 |
|---|---|---|---|---|---|---|---|---|---|---|---|---|---|---|---|---|---|---|---|---|
| | | | | | | | | | | | Year | | | | | | | | | |
| **With JFM** | | | | | | | | | | | | | | | | | | | | |
| FD | Revenue | 0 | 8,952 | 1,092,144 | 0 | 0 | 8,952 | 1,107,245 | 0 | 0 | 1,010,667 | 1,092,144 | 0 | 0 | 8,952 | 1,092,144 | 0 | 0 | 8,952 | 1,092,144 |
| Vill | Revenue | 0 | 8,952 | 1,092,144 | 0 | 0 | 8,952 | 1,107,245 | 0 | 0 | 1,010,667 | 1,092,144 | 0 | 0 | 8,952 | 1,092,144 | 0 | 0 | 8,952 | 1,092,144 |
| | Headloaders | 21,785 | 21,697 | 21,616 | 21,541 | 21,472 | 21,407 | 21,348 | 21,293 | 21,243 | 21,196 | 21,885 | 21,184 | 20,836 | 20,820 | 20,805 | 20,791 | 20,778 | 20,766 | 20,755 |
| | Graziers | 29,173 | 27,845 | 26,603 | 25,442 | 24,357 | 23,346 | 22,402 | 21,523 | 20,705 | 19,943 | 54,617 | 54,304 | 54,013 | 53,742 | 53,491 | 53,257 | 53,041 | 52,840 | 52,653 |
| | Collectors | 446,600 | 446,600 | 446,600 | 585,350 | 647,850 | 647,850 | 647,850 | 647,850 | 786,600 | 849,100 | 830,600 | 275,600 | 25,600 | 25,600 | 25,600 | 24,600 | 24,600 | 24,600 | 23,600 |
| | Total | 497,557 | 505,094 | 1,586,963 | 632,333 | 693,679 | 701,555 | 1,798,846 | 690,667 | 828,548 | 1,900,906 | 1,999,246 | 351,088 | 100,449 | 109,114 | 1,192,039 | 98,648 | 98,418 | 107,157 | 1,189,152 |
| | Non-revenue | 497,557 | 496,142 | 494,819 | 632,333 | 693,679 | 692,603 | 691,601 | 690,667 | 828,548 | 890,239 | 907,102 | 351,088 | 100,449 | 100,162 | 99,895 | 98,648 | 98,418 | 98,205 | 97,008 |
| Economic | | 497,557 | 514,046 | 2,679,107 | 632,333 | 693,679 | 710,507 | 2,906,091 | 690,667 | 828,548 | 2,911,572 | 3,091,390 | 351,088 | 100,449 | 118,066 | 2,284,183 | 98,648 | 98,418 | 116,109 | 2,281,296 |
| **Without JFM** | | | | | | | | | | | | | | | | | | | | |
| FD | Revenue | 266,490 | 266,490 | 266,490 | 266,490 | 266,490 | 266,490 | 266,490 | 266,490 | 266,490 | 266,490 | 266,490 | 266,490 | 266,490 | 266,490 | 266,490 | 266,490 | 266,490 | 266,490 | 266,490 |
| Vil | Revenue | 0 | 0 | 0 | 0 | 0 | 0 | 0 | 0 | 0 | 0 | 0 | 0 | 0 | 0 | 0 | 0 | 0 | 0 | 0 |
| | Headloaders | 9,695 | 9,477 | 9,269 | 9,068 | 8,875 | 8,689 | 8,510 | 8,338 | 8,173 | 8,013 | 7,859 | 7,711 | 7,568 | 7,430 | 7,296 | 7,167 | 7,043 | 6,923 | 6,807 |
| | Graziers | 70,716 | 70,100 | 69,507 | 68,935 | 68,384 | 67,854 | 67,344 | 66,852 | 66,379 | 65,924 | 65,487 | 65,066 | 64,661 | 64,271 | 63,897 | 63,536 | 63,190 | 62,857 | 62,536 |
| | Collectors | 37,000 | 37,000 | 37,000 | 39,000 | 39,000 | 39,000 | 39,000 | 39,000 | 39,000 | 39,000 | 39,000 | 39,000 | 41,000 | 41,000 | 41,000 | 41,000 | 41,000 | 41,000 | 41,000 |
| | Total village | 117,411 | 116,578 | 115,776 | 117,003 | 116,259 | 115,543 | 114,854 | 114,191 | 113,552 | 112,937 | 112,346 | 111,777 | 113,228 | 112,701 | 112,193 | 111,704 | 111,233 | 110,779 | 110,342 |
| | Non-revenue | 117,411 | 116,578 | 115,776 | 117,003 | 116,259 | 115,543 | 114,854 | 114,191 | 113,552 | 112,937 | 112,346 | 111,777 | 113,228 | 112,701 | 112,193 | 111,704 | 111,233 | 110,779 | 110,342 |
| Economic | | 383,901 | 383,068 | 382,266 | 383,493 | 382,749 | 382,033 | 381,344 | 380,681 | 380,042 | 379,427 | 378,836 | 378,267 | 379,718 | 379,191 | 378,683 | 378,194 | 377,723 | 377,269 | 376,832 |
| **Incremental benefits due to JFM** | | | | | | | | | | | | | | | | | | | | |
| FD | Revenue | -266,490 | -257,538 | 825,654 | -266,490 | -266,490 | -257,538 | 840,755 | -266,490 | -266,490 | 744,177 | 825,654 | -266,490 | -266,490 | -257,538 | 825,654 | -266,490 | -266,490 | -257,538 | 825,654 |
| Vil | Revenue | 0 | 8,952 | 1,092,144 | 0 | 0 | 8,952 | 1,107,245 | 0 | 0 | 1,010,667 | 1,092,144 | 0 | 0 | 8,952 | 1,092,144 | 0 | 0 | 8,952 | 1,092,144 |
| | Headloaders | 12,090 | 12,220 | 12,347 | 12,473 | 12,597 | 12,718 | 12,838 | 12,955 | 13,070 | 13,183 | 14,026 | 13,473 | 13,269 | 13,390 | 13,509 | 13,623 | 13,735 | 13,843 | 13,948 |
| | Graziers | -41,543 | -42,255 | -42,904 | -43,493 | -44,027 | -44,508 | -44,941 | -45,329 | -45,674 | -45,981 | -10,870 | -10,762 | -10,648 | -10,529 | -10,406 | -10,279 | -10,149 | -10,017 | -9,883 |
| | Collectors | 409,600 | 409,600 | 409,600 | 546,350 | 608,850 | 608,850 | 608,850 | 608,850 | 747,600 | 810,100 | 791,600 | 236,600 | -15,400 | -15,400 | -15,400 | -16,400 | -16,400 | -16,400 | -17,400 |
| | Total village | 380,147 | 388,516 | 1,471,187 | 515,330 | 577,420 | 586,012 | 1,683,992 | 576,476 | 714,996 | 1,787,968 | 1,886,900 | 239,311 | -12,780 | -3,587 | 1,079,847 | -13,056 | -12,815 | -3,622 | 1,078,809 |
| | Non-revenue | 380,147 | 379,564 | 379,043 | 515,330 | 577,420 | 577,060 | 576,747 | 576,476 | 714,996 | 777,301 | 794,756 | 239,311 | -12,780 | -12,539 | -12,297 | -13,056 | -12,815 | -12,574 | -13,335 |
| Economic | | 113,657 | 130,978 | 2,296,841 | 248,840 | 310,930 | 328,474 | 2,524,747 | 309,986 | 448,506 | 2,532,145 | 2,712,554 | -27,179 | -279,270 | -261,125 | 1,905,501 | -279,546 | -279,305 | -261,160 | 1,904,463 |

**Table MT.19    (C) Years 35-50        Incremental Benefits to Village from Natural Forest and Plantation**

**Management**

| | | 35 | 36 | 37 | 38 | 39 | 40 | 41 | 42 | 43 | 44 | 45 | 46 | 47 | 48 | 49 | 50 |
|---|---|---|---|---|---|---|---|---|---|---|---|---|---|---|---|---|---|
| **With JFM** | | | | | | | | | | | | | | | | | |
| FD | Revenue | 2,976,171 | 0 | 8,952 | 340,176 | 0 | 0 | 8,952 | 340,176 | 0 | 0 | 8,952 | 367,254 | 155,016 | 0 | 8,952 | 469,356 |
| Vill | Revenue | 2,976,171 | 0 | 8,952 | 340,176 | 0 | 0 | 8,952 | 340,176 | 0 | 0 | 8,952 | 367,254 | 155,016 | 0 | 8,952 | 469,356 |
| | Headloaders | 20,745 | 20,618 | 20,618 | 20,618 | 20,618 | 20,618 | 20,618 | 20,618 | 20,618 | 20,618 | 20,618 | 20,618 | 20,618 | 20,618 | 20,618 | 20,618 |
| | Graziers | 52,480 | 24,600 | 24,600 | 24,600 | 24,600 | 24,600 | 24,600 | 24,600 | 24,600 | 24,600 | 24,600 | 24,600 | 24,600 | 24,600 | 24,600 | 24,600 |
| | Collectors | 23,600 | 0 | 13,875 | 13,875 | 20,125 | 20,125 | 145,000 | 145,000 | 145,000 | 201,250 | 201,250 | 340,000 | 340,000 | 402,500 | 402,500 | 402,500 |
| | Total | 3,072,995 | 45,218 | 68,045 | 399,269 | 65,343 | 65,343 | 199,170 | 530,394 | 190,218 | 246,468 | 255,420 | 752,472 | 540,234 | 447,718 | 456,670 | 917,074 |
| | Non-revenue | 96,824 | 45,218 | 59,093 | 59,093 | 65,343 | 65,343 | 190,218 | 190,218 | 190,218 | 246,468 | 246,468 | 385,218 | 385,218 | 447,718 | 447,718 | 447,718 |
| Economic | | 6,049,167 | 45,218 | 76,997 | 739,445 | 65,343 | 65,343 | 208,122 | 870,570 | 190,218 | 246,468 | 264,372 | 1,119,726 | 695,250 | 447,718 | 465,622 | 1,386,430 |
| **Without JFM** | | | | | | | | | | | | | | | | | |
| FD | Revenue | 4,407,293 | 202,335 | 202,335 | 202,335 | 202,335 | 202,335 | 202,335 | 202,335 | 202,335 | 202,335 | 202,335 | 202,335 | 202,335 | 202,335 | 202,335 | 202,335 |
| Vil | Revenue | 0 | 0 | 0 | 0 | 0 | 0 | 0 | 0 | 0 | 0 | 0 | 0 | 0 | 0 | 0 | 0 |
| | Headloaders | 6,694 | 0 | 0 | 0 | 0 | 0 | 0 | 0 | 0 | 0 | 0 | 0 | 0 | 0 | 0 | 0 |
| | Graziers | 62,227 | 58,677 | 58,677 | 58,677 | 58,677 | 58,677 | 58,677 | 58,677 | 58,677 | 58,677 | 58,677 | 58,677 | 58,677 | 58,677 | 58,677 | 58,677 |
| | Collectors | 41,000 | 12,300 | 12,300 | 12,300 | 12,300 | 12,300 | 12,300 | 12,300 | 12,300 | 12,300 | 12,300 | 12,300 | 12,300 | 12,300 | 12,300 | 12,300 |
| | Total village | 109,921 | 70,977 | 70,977 | 70,977 | 70,977 | 70,977 | 70,977 | 70,977 | 70,977 | 70,977 | 70,977 | 70,977 | 70,977 | 70,977 | 70,977 | 70,977 |
| | Non-revenue | 109,921 | 70,977 | 70,977 | 70,977 | 70,977 | 70,977 | 70,977 | 70,977 | 70,977 | 70,977 | 70,977 | 70,977 | 70,977 | 70,977 | 70,977 | 70,977 |
| Economic | | 4,517,215 | 273,312 | 273,312 | 273,312 | 273,312 | 273,312 | 273,312 | 273,312 | 273,312 | 273,312 | 273,312 | 273,312 | 273,312 | 273,312 | 273,312 | 273,312 |
| **Incremental benefits due to JFM** | | | | | | | | | | | | | | | | | |
| FD | Revenue | -1,431,122 | -202,335 | -193,383 | 137,841 | -202,335 | -202,335 | -193,383 | 137,841 | -202,335 | -202,335 | -193,383 | 164,919 | -47,319 | -202,335 | -193,383 | 267,021 |
| Vil | Revenue | 2,976,171 | 0 | 8,952 | 340,176 | 0 | 0 | 8,952 | 340,176 | 0 | 0 | 8,952 | 367,254 | 155,016 | 0 | 8,952 | 469,356 |
| | Headloaders | 14,050 | 20,618 | 20,618 | 20,618 | 20,618 | 20,618 | 20,618 | 20,618 | 20,618 | 20,618 | 20,618 | 20,618 | 20,618 | 20,618 | 20,618 | 20,618 |
| | Graziers | -9,747 | -34,077 | -34,077 | -34,077 | -34,077 | -34,077 | -34,077 | -34,077 | -34,077 | -34,077 | -34,077 | -34,077 | -34,077 | -34,077 | -34,077 | -34,077 |
| | Collectors | -17,400 | -12,300 | 1,575 | 1,575 | 7,825 | 7,825 | 132,700 | 132,700 | 132,700 | 188,950 | 188,950 | 327,700 | 327,700 | 390,200 | 390,200 | 390,200 |
| | Total village | 2,963,074 | -25,758 | -2,931 | 328,293 | -5,633 | -5,633 | 128,194 | 459,418 | 119,242 | 175,492 | 184,444 | 681,496 | 469,258 | 376,742 | 385,694 | 846,098 |
| | Non-revenue | -13,097 | -25,758 | -11,883 | -11,883 | -5,633 | -5,633 | 119,242 | 119,242 | 119,242 | 175,492 | 175,492 | 314,242 | 314,242 | 376,742 | 376,742 | 376,742 |
| Economic | | 1,531,952 | -228,093 | -196,314 | 466,134 | -207,968 | -207,968 | -65,189 | 597,259 | -83,093 | -26,843 | -8,939 | 846,414 | 421,939 | 174,407 | 192,311 | 1,113,119 |

101

**Table SC.1**          **Forest Area**

| Forest Type | Area (ha) |
|---|---|
| Natural forest | |
|     Undegraded | 30 |
|     Partially degraded | 20 |
|     Highly degraded | 23 |
|     Sub-total | 73 |
| Plantations | |
|     Eucalyptus | 20 |
|     Acacia | 30 |
|     Sub-total | 50 |
| Encroached land | 7 |
| Total | 130 |

**Table SC.2      Sample plots**

| Type | Degradation Status | No | Species | Sample Area (ha) |
|------|------|------|------|------|
| Natural Forest | Moderate | NF-1 | | 30 |
| Natural Forest | Partial | NF-2 | | 20 |
| Natural Forest | Highly | NF-3 | | 23 |
| Sub-total | | | | 73 |
| Plantation | None | PL-1 | Eucalyptus | 20 |
| Plantation | None | PL-2 | Acacia | 30 |
| Sub-total | | | | 50 |
| Total | | | | 123 |

**Table SC.3.  Silvicultural Parameters for Natural Forest**

| Sample Area | Area (ha) | Protection | Species | Stems (No/ha) | Basal Area (m^2/ha) | Height (m) | Form Factor | Growing Stock (m^3/ha) | Age (yrs) | MAI (m^3/ha/yr |
|---|---|---|---|---|---|---|---|---|---|---|
| NF.1 | 30 | Moderate | | 1,300 | 15 | 12.0 | 0.420 | 75.60 | 16 | 4.73 |
| NF.2 | 20 | Partial | | 1,400 | 6 | 7.0 | 0.420 | 17.64 | 10 | 1.76 |
| NF.3 | 23 | Highly | | 1,500 | 4 | 3.0 | 0.420 | 5.04 | 10 | 0.50 |
| PL.1 | 20 | Good | Eucalyptus | 1,500 | | | 0.420 | 20.00 | 10 | 2.00 |
| PL.2 | 30 | Good | Acacia | 1,800 | 9 | | 0.420 | 60.00 | 10 | 6.00 |
| Encroache | 7 | | | | | | | | | |

Note:     Study measurements

Growing stock (species) = Proportion of stems x Height x Basal Area x Form Factor

MAI = Growing Stock/Age

**Table SC.4**  **Forest Productivity Model (Natural Forest Type NF-1)**

| Management | Output | Unit | Unit Price (Rs/unit) | 0 | 1 | 2 | 3 | 4 | 5 | 6 | 7 | 8 | 9 | 10 |
|---|---|---|---|---|---|---|---|---|---|---|---|---|---|---|
| | | | | Year in Coppice Cycle | | | | | | | | | | |
| **With JFM** | | | | | | | | | | | | | | |
| Base coppice | Biomass (Total | m^3/ha | | 75.6 | | | | | | | | | | |
| | Poles (16 years | poles | 2,204.21 | 45.36 | | | | | | | | | | |
| | Fuelwood | m^3/ha | 276.50 | 30.24 | | | | | | | | | | |
| MMS-1 | Fuelwood | m^3/ha | 276.50 | | | | | 7.04 | | | | | | |
| MSS-2 | Fuelwood | m^3/ha | 276.50 | | | | | | | 5.33 | | | | |
| | Poles ( 7 years) | poles | 1,211.01 | | | | | | | 5.33 | | | | |
| Final coppice | Biomass | m^3/ha | | | | | | | | | | | | 50.33 |
| | Poles (7/10 yrs) | poles | 1,476.76 | | | | | | | | | | | 30.20 |
| | Fuelwood | m^3/ha | 276.50 | | | | | | | | | | | 20.13 |
| Growing stock | GS | m^3/ha | | 75.6 | | | | | | | | | | |
| Canopy cover | CC | (%) | | 100 | 10 | 30 | 60 | 80 | 33 | 66 | 80 | 75 | 90 | 100 |
| NTFP collection | | | | | | | | | | | | | | |
| | Mushrooms | kg | 9.33 | 211 | 21 | 63 | 127 | 169 | 70 | 139 | 169 | 158 | 190 | 211 |
| | Tendu | bag | 130.00 | 0 | 0 | 0 | 0 | 0 | 0 | 0 | 0 | 0 | 0 | 0 |
| | Other | LS | 698.31 | 1 | 1 | 1 | 1 | 1 | 1 | 1 | 1 | 1 | 1 | 1 |
| **Without JFM** | | | | | | | | | | | | | | |
| GS (standing) | | m^3/ha | 1,433.13 | 75.60 | 75.60 | 80.66 | 85.73 | 91.13 | 96.87 | 102.98 | 109.47 | 116.37 | 123.70 | 131.49 |
| GS (increment) | | m^3/ha | 1,890.70 | | | | | | | | | | | 55.89 |
| CAI | | m^3/ha/yr | | 5.63 | 5.63 | 5.63 | 6.00 | 6.38 | 6.78 | 7.21 | 7.66 | 8.15 | 8.66 | 9.21 |
| Lopping/hacking (% CAI) | | 10 | 276.50 | 0.56 | 0.56 | 0.56 | 0.60 | 0.64 | 0.68 | 0.72 | 0.77 | 0.81 | 0.87 | 0.92 |
| MAI (standing) | | m^3/ha/yr | | 4.73 | 4.45 | 4.48 | 4.51 | 4.56 | 4.61 | 4.68 | 4.76 | 4.85 | 4.95 | 5.06 |
| GS (total) | | m^3/ha | | 75.6 | | | | | | | | | | 138.62 |
| MAI (total) | | m^3/ha/yr | | | | | | | | | | | | 5.33 |
| Canopy cover | CC (%) | | | | | | | | | | | | | |
| NTFP collection | | | | | | | | | | | | | | |
| | Mushrooms | kg | 9.33 | 211 | 211 | 222 | 232 | 244 | 256 | 269 | 282 | 297 | 313 | 329 |
| | Tendu | bag | 130.00 | 0 | 0 | 0 | 0 | 0 | 0 | 0 | 0 | 0 | 0 | 0 |
| | Other | LS | 698 | 1 | 1 | 1 | 1 | 1 | 1 | 1 | 1 | 1 | 1 | 1 |

**Table SC.5**            **Forest Productivity Model (Natural Forest Type NF-2)**

| Management | Output | Unit | Unit Price (Rs/unit) | Year in Coppice Cycle | | | | | | | | | | |
|---|---|---|---|---|---|---|---|---|---|---|---|---|---|---|
| | | | | 0 | 1 | 2 | 3 | 4 | 5 | 6 | 7 | 8 | 9 | 10 |
| **With JFM** | | | | | | | | | | | | | | |
| Base coppice | Biomass (Tota | m^3/ha | | 17.64 | | | | | | | | | | |
| | Poles (10 year | poles | 1,609.64 | 10.584 | | | | | | | | | | |
| | Fuelwood | m^3/h | 276.50 | 7.056 | | | | | | | | | | |
| MMS-1 | Fuelwood | m^3/h | 276.50 | | | | | 7.04 | | | | | | |
| MSS-2 | Fuelwood | m^3/h | 276.50 | | | | | | | | 5.33 | | | |
| | Poles ( 7 years | poles | 1,211.01 | | | | | | | | 5.33 | | | |
| Final coppice | Biomass | m^3/ha | | | | | | | | | | | | 50.33 |
| | Poles (7/10 yrs | poles | 1,476.76 | | | | | | | | | | | 30.20 |
| | Fuelwood | m^3/h | 276.50 | | | | | | | | | | | 20.13 |
| Growing stock | GS | m^3/ha | | 17.64 | | | | | | | | | | |
| Canopy cover | CC (%) | (%) | | | 10 | 30 | 60 | 80 | 33 | 66 | 80 | 75 | 90 | 100 |
| NTFP collection | | | | | | | | | | | | | | |
| | Mushrooms | kg | 9.33 | 89 | 9 | 27 | 53 | 71 | 29 | 59 | 71 | 67 | 80 | 89 |
| | Tendu | bag | 130.00 | 0.89 | 0.89 | 0.89 | 0.89 | 0.89 | 0.89 | 0.89 | 0.89 | 0.89 | 0.89 | 0.89 |
| | Other | LS | 698.31 | 1 | 1 | 1 | 1 | 1 | 1 | 1 | 1 | 1 | 1 | 1 |
| **Without JFM** | | | | | | | | | | | | | | |
| | GS (standing) | m^3/h | 1,076.38 | 17.64 | 17.64 | 18.30 | 18.95 | 19.64 | 20.34 | 21.07 | 21.83 | 22.62 | 23.43 | 24.27 |
| | GS (increment) | m^3/h | 1,076.38 | | | | | | | | | | | 6.63 |
| | CAI | m^3/ha/yr | | 1.31 | 1.31 | 1.31 | 1.36 | 1.41 | 1.46 | 1.52 | 1.57 | 1.63 | 1.69 | 1.75 |
| Lopping/hacking (% CAI) | | 50 | 276.50 | 0.66 | 0.66 | 0.66 | 0.68 | 0.71 | 0.73 | 0.76 | 0.79 | 0.81 | 0.84 | 0.87 |
| | MAI (standing) | m^3/ha/yr | | 1.10 | 1.04 | 1.02 | 1.00 | 0.98 | 0.97 | 0.96 | 0.95 | 0.94 | 0.94 | 0.93 |
| | GS (total) | m^3/ha | | 17.64 | | | | | | | | | | 31.78 |
| | MAI (total) | m^3/ha/yr | | | | | | | | | | | | 1.22 |
| Canopy cover | CC (%) | | | | | | | | | | | | | |
| NTFP collection | | | | | | | | | | | | | | |
| | Mushrooms | kg | 9.33 | 89 | 89 | 90 | 91 | 93 | 94 | 96 | 98 | 99 | 101 | 103 |
| | Tendu | bag | 130.00 | 0.89 | 0.89 | 0.88 | 0.87 | 0.86 | 0.86 | 0.85 | 0.84 | 0.83 | 0.82 | 0.80 |
| | Other | LS | 698.31 | 1 | 1 | 1 | 1 | 1 | 1 | 1 | 1 | 1 | 1 | 1 |

**Incremental Benefits due to JFM**

| | Net Worth (Rs) | Discount Rate (%) | | 0 | 1 | 2 | 3 | 4 | 5 | 6 | 7 | 8 | 9 | 10 |
|---|---|---|---|---|---|---|---|---|---|---|---|---|---|---|
| **Wood products** | | | | | | | | | | | | | | |
| With | 35,679 | 0.12 | | 18,987 | 0 | 0 | 0 | 1,945 | 0 | 0 | 7,928 | 0 | 0 | 50,162 |
| Without | 3,230 | 0.12 | | 182 | 182 | 182 | 189 | 195 | 202 | 210 | 217 | 225 | 233 | 7,382 |
| Increment | 32,449 | 0.12 | | 18,806 | -182 | -182 | -189 | 1,750 | -202 | -210 | 7,711 | -225 | -233 | 42,780 |
| **NTFP** | | | | | | | | | | | | | | |
| With | 7,865 | 0.12 | | 1,645 | 897 | 1,063 | 1,311 | 1,477 | 1,087 | 1,361 | 1,477 | 1,435 | 1,559 | 1,642 |
| Without | 9,980 | 0.12 | | 1,644 | 1,642 | 1,654 | 1,666 | 1,678 | 1,691 | 1,704 | 1,718 | 1,732 | 1,746 | 1,762 |
| Increment | -2,115 | 0.12 | | 0 | -745 | -591 | -355 | -202 | -603 | -343 | -241 | -297 | -187 | -119 |
| **Total** | | | | | | | | | | | | | | |
| With | 43,545 | 0.12 | | 20,632 | 897 | 1,063 | 1,311 | 3,422 | 1,087 | 1,361 | 9,405 | 1,435 | 1,559 | 51,804 |
| Without | 13,210 | 0.12 | | 1,826 | 1,824 | 1,836 | 1,854 | 1,873 | 1,893 | 1,914 | 1,935 | 1,957 | 1,979 | 9,143 |
| Increment | 30,334 | 0.12 | | 18,806 | -927 | -773 | -543 | 1,548 | -806 | -553 | 7,470 | -522 | -420 | 42,661 |

**Table SC.6**         **Forest Productivity Model (Natural Forest Type NF-3)**

| Management | Output | Unit | Unit Prices (Rs/unit) | 0 | 1 | 2 | 3 | 4 | 5 | 6 | 7 | 8 | 9 | 10 |
|---|---|---|---|---|---|---|---|---|---|---|---|---|---|---|
| | | | | | | | | Year in Coppice Cycle | | | | | | |
| **With JFM** | | | | | | | | | | | | | | |
| Base coppice | Biomass (Total | m^3/ha | | 5.04 | | | | | | | | | | |
| | Poles (10 years | poles | 1,609.64 | 3.024 | | | | | | | | | | |
| | Fuelwood | m^3/ha | 276.50 | 2.016 | | | | | | | | | | |
| MMS-1 | Fuelwood | m^3/ha | 276.50 | | | | | 7.04 | | | | | | |
| MSS-2 | Fuelwood | m^3/ha | 276.50 | | | | | | | | 5.33 | | | |
| | Poles ( 7 years) | poles | 1,211.01 | | | | | | | | 5.33 | | | |
| Final coppice | Biomass | m^3/ha | | | | | | | | | | | | 50.33 |
| | Poles (7/10 yrs) | poles | 1,476.76 | | | | | | | | | | | 30.20 |
| | Fuelwood | m^3/ha | 276.50 | | | | | | | | | | | 20.13 |
| Growing stock | GS | m^3/ha | | 5.04 | | | | | | | | | | |
| Canopy cover | CC (%) | (%) | | 10 | 10 | 30 | 60 | 80 | 33 | 66 | 80 | 75 | 90 | 100 |
| NTFP collection | | | | | | | | | | | | | | |
| | Mushrooms | kg | 9.33 | 6 | 6 | 19 | 37 | 50 | 21 | 41 | 50 | 47 | 56 | 62 |
| | Tendu | bag | 130.00 | 1.05 | 1.05 | 1.05 | 1.05 | 1.05 | 1.05 | 1.05 | 1.05 | 1.05 | 1.05 | 1.05 |
| | Other | LS | 698.31 | 1 | 1 | 1 | 1 | 1 | 1 | 1 | 1 | 1 | 1 | 1 |
| **Without JFM** | | | | | | | | | | | | | | |
| GS (standing) | | m^3/ha | 1,076.38 | 5.04 | 5.04 | 5.04 | 5.04 | 5.04 | 5.04 | 5.04 | 5.04 | 5.04 | 5.04 | 5.04 |
| GS (increment) | | m^3/ha | 276.50 | | | | | | | | | | | 0.00 |
| CAI | | m^3/ha/yr | | 0.38 | 0.38 | 0.38 | 0.38 | 0.38 | 0.38 | 0.38 | 0.38 | 0.38 | 0.38 | 0.38 |
| Lopping/hacking (% CAI) | | 100 | 276.50 | 0.38 | 0.38 | 0.38 | 0.38 | 0.38 | 0.38 | 0.38 | 0.38 | 0.38 | 0.38 | 0.38 |
| MAI Standing | | m^3/ha/yr | | 0.32 | 0.30 | 0.28 | 0.27 | 0.25 | 0.24 | 0.23 | 0.22 | 0.21 | 0.20 | 0.19 |
| GS (total) | | m^3/ha | | 5.04 | | | | | | | | | | 8.81 |
| MAI (total) | | m^3/ha/yr | | | | | | | | | | | | 0.34 |
| Canopy cover | CC (%) | | | | | | | | | | | | | |
| NTFP collection | | | | | | | | | | | | | | |
| | Mushrooms | kg | 9.33 | 62 | 62 | 62 | 62 | 62 | 62 | 62 | 62 | 62 | 62 | 62 |
| | Tendu | bag | 130.00 | 1.05 | 1.05 | 1.05 | 1.05 | 1.05 | 1.05 | 1.05 | 1.05 | 1.05 | 1.05 | 1.05 |
| | Other | LS | 698.31 | 1 | 1 | 1 | 1 | 1 | 1 | 1 | 1 | 1 | 1 | 1 |

**Incremental Benefits due to JFM**

| | Net Worth (Rs) | Discount Rate (%) | | 0 | 1 | 2 | 3 | 4 | 5 | 6 | 7 | 8 | 9 | 10 |
|---|---|---|---|---|---|---|---|---|---|---|---|---|---|---|
| **Wood products** | | | | | | | | | | | | | | |
| With | 23,570 | 0.12 | | 5,425 | 0 | 0 | 0 | 1,945 | 0 | 0 | 7,928 | 0 | 0 | 50,162 |
| Without | 620 | 0.12 | | 104 | 104 | 104 | 104 | 104 | 104 | 104 | 104 | 104 | 104 | 104 |
| Increment | 22,951 | 0.12 | | 5,321 | -104 | -104 | -104 | 1,841 | -104 | -104 | 7,824 | -104 | -104 | 50,057 |
| **NTFP** | | | | | | | | | | | | | | |
| With | 6,616 | 0.12 | | 893 | 893 | 1,009 | 1,183 | 1,299 | 1,027 | 1,218 | 1,299 | 1,270 | 1,357 | 1,415 |
| Without | 8,404 | 0.12 | | 1,415 | 1,415 | 1,415 | 1,415 | 1,415 | 1,415 | 1,415 | 1,415 | 1,415 | 1,415 | 1,415 |
| Increment | -1,788 | 0.12 | | -522 | -522 | -406 | -232 | -116 | -389 | -197 | -116 | -145 | -58 | 0 |
| **Total** | | | | | | | | | | | | | | |
| With | 30,186 | 0.12 | | 6,318 | 893 | 1,009 | 1,183 | 3,245 | 1,027 | 1,218 | 9,228 | 1,270 | 1,357 | 51,577 |
| Without | 9,023 | 0.12 | | 1,520 | 1,520 | 1,520 | 1,520 | 1,520 | 1,520 | 1,520 | 1,520 | 1,520 | 1,520 | 1,520 |
| Increment | 21,162 | 0.12 | | 4,799 | -626 | -510 | -336 | 1,725 | -493 | -302 | 7,708 | -249 | -162 | 50,057 |
| With | 6,126,100 | 0.12 | | 4,239,292 | 97,307 | 120,189 | 154,511 | 375,128 | 123,621 | 161,375 | 867,244 | 171,672 | 188,833 | 4,376,869 |
| Without | 1,972,893 | 0.12 | | 215,234 | 141,492 | 146,450 | 154,518 | 160,442 | 161,740 | 163,109 | 164,554 | 166,079 | 167,689 | 3,612,049 |
| Increment | 4,153,207 | 0.12 | | 4,024,058 | -44,185 | -26,261 | -7 | 214,686 | -38,119 | -1,734 | 702,690 | 5,593 | 21,144 | 764,820 |

**Table SC.7**  **Plantation Costs and Benefits (Eucalyptus/Acacia)**

| | Unit | Rate (Rs/unit) | Year 0 | 1 | 2-5 | 6 | 7 | 8-10 | 11-13 | 14 | 15-20 | 21 | 22-27 | 28 |
|---|---|---|---|---|---|---|---|---|---|---|---|---|---|---|
| **Inputs and costs** | | | | | | | | | | | | | | |
| **Labour** | | | | | | | | | | | | | | |
| Nursery operations | Day | 28 | 10 | 25 | | | | | | | | | | |
| Survey, cleaning, allignment | Day | 28 | 5 | | | | | | | | | | | |
| Digging 'V' trench | Day | 28 | 10 | | | | | | | | | | | |
| Digging pits | Day | 28 | 15 | 30 | | | | | | | | | | |
| Filling pits | Day | 28 | | 10 | | | | | | | | | | |
| Planting with potted seedling | Day | 28 | | 15 | | | | | | | | | | |
| Weeding | Day | 28 | | 25 | | | | | | | | | | |
| Watch and ward | Year | 3600 | | 0.03 | 0.03 | | | | | | | | | |
| Sub-total | | | 1120 | 3060 | 120 | | | | | | | | | |
| **Materials** | | | | | | | | | | | | | | |
| Shed and fencing | LS | | 40 | 20 | | | | | | | | | | |
| FYM | LS | | 80 | | | | | | | | | | | |
| Seeds | LS | | 20 | 10 | | | | | | | | | | |
| Watering cans etc | LS | | 10 | | | | | | | | | | | |
| Misc | LS | | 10 | | | | | | | | | | | |
| Sub-total | | | 160 | 30 | 0 | | | | | | | | | |
| **Total** | | | 1280 | 3090 | 120 | | | | | | | | | |
| **Outputs and benefits** | | | | | | | | | | | | | | |
| Poles/fuelwood | m^3/ha | 369.30 | | | | | 28 | | | 28 | | 28 | | 28 |
| | Rs/ha | | 0 | 0 | 0 | 0 | 10,340 | 0 | 0 | 10,340 | 0 | 10,340 | 0 | 10,340 |
| Tendu | std bag/ha | 130 | 0.26 | 0.26 | 0.26 | 0.26 | 0.26 | 0.26 | 0 | 0 | 0 | 0 | 0 | 0 |
| | Rs/ha | | 34 | 34 | 34 | 34 | 34 | 34 | 0 | 0 | 0 | 0 | 0 | 0 |
| **Total** | Rs/ha | | 34 | 34 | 34 | 34 | 10,375 | 34 | 0 | 10,340 | 0 | 10,340 | 0 | 10,340 |
| | NPV(12%,28 yrs) | | | | | | | | | | | | | |
| Net benefit | 3,613 | | -1,246 | -3,056 | -86 | 34 | 10,375 | 34 | 0 | 10,340 | 0 | 10,340 | 0 | 10,340 |
| **Without plantation and JFM** | | | | | | | | | | | | | | |
| Loss of Tendu (% of NF-3) | | 0.50 | 69 | 69 | 69 | 69 | 69 | 69 | 0 | 0 | 0 | 0 | 0 | 0 |
| Total | | | 69 | 69 | 69 | 69 | 69 | 69 | 0 | 0 | 0 | 0 | 0 | 0 |
| | NPV(12%,28 yrs) | | | | | | | | | | | | | |
| Incremental benefit (Rs/ha) | 3,206 | | -1,314 | -3,124 | -154 | -34 | 10,306 | -34 | 0 | 10,340 | 0 | 10,340 | 0 | 10,340 |
| | NPV(12%,28 yrs) | | | | | | | | | | | | | |
| Incremental benefit (Rs per 50 ha) | 160,315 | | -65,714 | -156,214 | -7,714 | -1,714 | 515,303 | -1,714 | 0 | 517,017 | 0 | 517,017 | 0 | 517,017 |

**Table SC.8    Sal Productivity**

| Year | Standing Volume (SV) (cm^3/ha) | Current Annual Increment (CAI) (cm^3/ha/yr) | Mean Annual Increment (MAI) (cm^3/ha/yr) | Current Increment Percent (CIP) (CAI/%GS) | Reciprocal of SV (1/SV) |
|---|---|---|---|---|---|
| 1 | | | | | |
| 2 | | | | | |
| 3 | 0.004 | 0.004 | 0.001 | 100.00 | 263.16 |
| 4 | 0.005 | 0.001 | 0.001 | 24.00 | 200.00 |
| 5 | 0.007 | 0.002 | 0.001 | 28.57 | 142.86 |
| 6 | 0.009 | 0.002 | 0.002 | 22.22 | 111.11 |
| 7 | 0.012 | 0.003 | 0.002 | 25.00 | 83.33 |
| 8 | 0.017 | 0.005 | 0.002 | 28.57 | 59.52 |
| 9 | 0.022 | 0.005 | 0.002 | 24.32 | 45.05 |
| 10 | 0.030 | 0.007 | 0.003 | 24.75 | 33.90 |
| 11 | 0.038 | 0.008 | 0.003 | 21.33 | 26.67 |
| 12 | 0.045 | 0.008 | 0.004 | 17.40 | 22.03 |
| 13 | 0.055 | 0.009 | 0.004 | 17.00 | 18.28 |
| 14 | 0.062 | 0.007 | 0.004 | 11.77 | 16.13 |
| 15 | 0.071 | 0.009 | 0.005 | 12.68 | 14.08 |
| 16 | 0.081 | 0.010 | 0.005 | 12.35 | 12.35 |
| 17 | 0.090 | 0.009 | 0.005 | 10.00 | 11.11 |
| 18 | 0.100 | 0.010 | 0.006 | 9.82 | 10.02 |
| 19 | 0.109 | 0.009 | 0.006 | 8.02 | 9.22 |
| 20 | 0.118 | 0.009 | 0.006 | 7.66 | 8.51 |
| 21 | 0.126 | 0.009 | 0.006 | 6.75 | 7.94 |
| 22 | 0.135 | 0.009 | 0.006 | 6.53 | 7.42 |
| 23 | 0.143 | 0.008 | 0.006 | 5.87 | 6.98 |
| 24 | 0.153 | 0.010 | 0.006 | 6.28 | 6.54 |
| 25 | 0.161 | 0.008 | 0.006 | 5.09 | 6.21 |

Source:      Standard yield tables

Note:        Growing Stock (GS) (m^3/ha) = 1000 x Stem volume (cm^3/ha)

$CIP\% = a + b/GS$

**Figure SC2. Sal Growth Rates**

Regression Output:

| | |
|---|---|
| Constant | 7.44 |
| Std Err of Y Est | 10.92 |
| R Squared | 0.70 |
| No. of Observations | 23 |
| Degrees of Freedom | 21 |
| | |
| X Coefficient(s) | 0.2359 |
| Std Err of Coef. | 0.0339 |

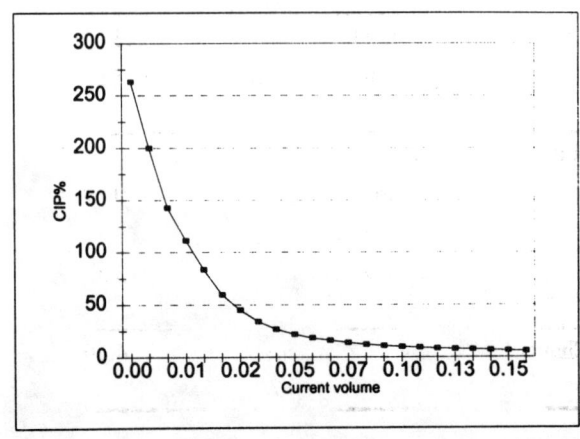

**Table SC.9. Estimated Offtake Rates**

| Degradation | Offtake rate %CAI |
|---|---|
| Moderate | 10 |
| Partial | 50 |
| High | 100 |

**Table SC.10    Projected Yields in Natural Forest**

| | Product | Unit | Rate | NF-1 | NF-2 | NF-3 | Growth | offtake |
|---|---|---|---|---|---|---|---|---|
| | | | | Sample areas | | | Biomass Balance | |
| **With JFM** | | | | | | | | |
| Initial coppice | | | | | | | | |
| | Biomass | m^3/ha | | 75.6 | 17.6 | 5 | | |
| | Fuelwood | m^3/ha | 0.4 | 30.24 | 7.04 | 2 | | |
| | Poles | No | 0.6 | 45.36 | 10.56 | 3 | | |
| | Poles | No | | 1300 | 1400 | 1500 | | |
| MSS-I | Biomass | m^3/ha | | 7.11 | 7.11 | 7.11 | 21.32 | 7.11 |
| MSS-II | Biomass | m^3/ha | | 5.33 | 5.33 | 5.33 | 30.20 | 5.33 |
| | Poles | No | 6 | 217 | 233 | 250 | | |
| | Biomass | m^3/ha | 0.022 | 4.77 | 5.13 | 5.50 | | 5.00 |
| Final cut | Biomass | m^3/ha | | 50.6 | 50.6 | 50.6 | 50.6 | 17.44 |
| | Fuelwood | m^3/ha | 0.4 | 20.24 | 20.24 | 20.24 | 68.04 | |
| | Poles | No | 0.6 | 30.36 | 30.36 | 30.36 | | |
| | Poles | Small | | | | | | |
| | | Medium | | | | | | |
| Total Biomass (rotatio | | m^3/ha | | 67.80 | 68.17 | 68.54 | | |
| MAI (rotatio | | m^3/ha/yr | | 6.78 | 6.82 | 6.85 | | |
| Total Biomass (to date) | | | | 143.40 | 85.77 | 73.54 | | |
| MAI (to date | | m^3/ha/yr | | 8.96 | 8.58 | 7.35 | | |
| **Without JFM** | | | | | | | | |
| | Total Bioma | m^3/ha | | 139.00 | 32.00 | 9.00 | | |
| | Hacking rate (%) | | | 10 | 50 | 100 | | |
| | MAI (to date | m^3/ha/yr | | 5.35 | 3.20 | 0.90 | | |
| Change in biomass | | m^3/ha | | 4.40 | 53.77 | 64.54 | | |

**Table SC.11**     **Mushroom Production**

| FPC | Forest Type | Basal Area (m^2) | Height (m^2) | Form Factor | Growing Stock (m^3) | Age (yr) | MAI (m^3/ha/y) | Area (ha) | Leaf litter Per ha (MT/ha) | Total (MT/ha) | Offtake (%) | Total retained | Mushrooms Total (kg) | Yield (kg/ha) |
|---|---|---|---|---|---|---|---|---|---|---|---|---|---|---|
| Fakirdunga | | | | | | | | | | | | | | |
| | NF-1 | 15 | 12 | 0.4 | 72 | 16 | 4.5 | 23 | 3.28 | 75.54 | 10 | 67.99 | 4,390 | 191 |
| | NF-2 | 6 | 7 | 0.4 | 17 | 10 | 1.7 | 20 | 2.19 | 43.75 | 40 | 26.25 | 1,695 | 85 |
| | NF-3 | 4 | 3 | 0.4 | 5 | 10 | 0.5 | 30 | 1.94 | 58.31 | 60 | 23.32 | 1,506 | 50 |
| | Overall | 8.0 | | | 29 | | | 73 | | 177.60 | | 117.56 | 7,590 | 104 |
| Shyampur | | | | | | | | | | | | | | |
| | NF-1 | 25 | 11 | 0.4 | 110 | 8 | 13.8 | 30 | 4.50 | 135.11 | 0 | 135.11 | 8,086 | 270 |
| | NF-2 | 15 | 8 | 0.4 | 48 | 4 | 12.0 | 60 | 3.28 | 197.07 | 5 | 187.22 | 11,204 | 187 |
| | NF-3 | 8 | 6 | 0.4 | 19 | 3 | 6.4 | 40 | 2.43 | 97.25 | 30 | 68.07 | 4,074 | 102 |
| | Overall | 15.2 | | | 53 | | | 130 | | 429.42 | | 390.40 | 23,364 | 180 |
| Total | | | | | | | | 203 | | | | | 30,954 | 152 |
| Barakulia | | | | | | | | | | | | | | |
| | NF-1 | 16 | 11 | 0.4 | 70 | 10 | 7.0 | 10 | 3.41 | 34.06 | | | | |
| | NF-2 | 14 | 10 | 0.4 | 56 | 10 | 5.6 | 30 | 3.16 | 94.88 | | | | |
| | NF-3 | | | | | | | | | | | | | |
| | Overall | | | | | | | | | 128.94 | | | | |

Relationships:    Leaf litter (LL) = 1.456 + 0.1219 x B    (Leaf litter in dry tons/ha/year)
                                                                                (Source: SPWD Manual Vol 2)

Total LL = LL x area

Mushrooms = 51.48 + 2.11 x GS    (Mushrooms in tons/ha/year)
                                                           (Source: Regression on field data - see below)

Mushroom production per ha is based on proportion of LL in each sample plot

Relationship between GS and Mushroom Production
            (based on selected data from field study)

Mushroom production = a + b x GS

Regression Output:
Constant                          51.48
Std Err of Y Est              11.85
R Squared                         0.99
No. of Observations            5
Degrees of Freedom            3

X Coefficient(s)            2.11174
Std Err of Coef.            0.13714

**Figure SC.2 Mushroom Production**

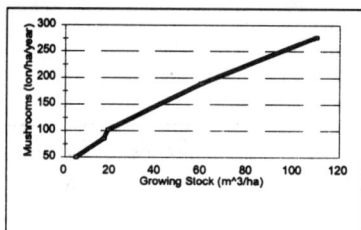

**Table SC.12**  **Incremental Benefits to Village from Natural Forest (73 ha)**

| Incremental benefits from JFM | Net Worth (Rs) | Discount Rate (%) | 0 | 1 | 2 | 3 | 4 | 5 | 6 | 7 | 8 | 9 | 10 |
|---|---|---|---|---|---|---|---|---|---|---|---|---|---|
| **Wood products** | | | | | | | | | | | | | |
| With JFM | 1,367,022 | 12 | 0 | 0 | 0 | 0 | 142,010 | 0 | 0 | 578,776 | 0 | 0 | 3,661,813 |
| Without JFM | 1,022,660 | 12 | 10,703 | 10,703 | 10,703 | 11,150 | 11,598 | 12,072 | 12,572 | 13,099 | 13,655 | 14,242 | 3,327,872 |
| Increment to JF | 344,362 | 12 | -10,703 | -10,703 | -10,703 | -11,150 | 130,412 | -12,072 | -12,572 | 565,677 | -13,655 | -14,242 | 333,941 |
| **NTFP** | | | | | | | | | | | | | |
| With JFM | 649,980 | 12 | 133,477 | 65,344 | 83,141 | 109,836 | 127,633 | 85,810 | 115,175 | 127,633 | 123,183 | 136,531 | 145,429 |
| Without JFM | 928,332 | 12 | 145,474 | 145,429 | 148,658 | 151,887 | 155,325 | 158,972 | 162,841 | 166,948 | 171,306 | 175,930 | 180,838 |
| Increment to JF | -278,351 | 12 | -11,997 | -80,085 | -65,517 | -42,051 | -27,692 | -73,161 | -47,667 | -39,315 | -48,122 | -39,399 | -35,409 |
| **Total** | | | | | | | | | | | | | |
| With JFM | 2,017,002 | 12 | 133,477 | 65,344 | 83,141 | 109,836 | 269,643 | 85,810 | 115,175 | 706,409 | 123,183 | 136,531 | 3,807,242 |
| Without JFM | 1,950,992 | 12 | 156,177 | 156,132 | 159,361 | 163,037 | 166,923 | 171,043 | 175,413 | 180,047 | 184,961 | 190,173 | 3,508,710 |
| Increment to JF | 66,011 | 12 | -22,700 | -90,788 | -76,220 | -53,201 | 102,720 | -85,233 | -60,238 | 526,362 | -61,778 | -53,642 | 298,531 |

**Table SC.13**  **Total Incremental Benefits to Village from Natural Forest and Plantations**

| Incremental Benefits | Net Worth (Rs, 32 yrs) | Discount Rate (%) | Rotation 1 Coppice Year | | | | | | | | | | |
|---|---|---|---|---|---|---|---|---|---|---|---|---|---|
| | | | 0 | 1 | 2 | 3 | 4 | 5 | 6 | 7 | 8 | 9 | 10 |
| **Natural Forest (73 ha)** | | | | | | | | | | | | | |
| **Costs** | | | | | | | | | | | | | |
| Operations (73 ha) | 110,564 | 12 | 19,511 | 8,760 | 8,760 | 26,826 | 8,760 | 8,760 | 8,760 | 36,939 | 8,760 | 8,760 | 0 |
| **Wood products** | | | | | | | | | | | | | |
| With JFM | 1,766,667 | 12 | 0 | 0 | 0 | 0 | 142,010 | 0 | 0 | 578,776 | 0 | 0 | 3,661,813 |
| Without JFM | 1,320,880 | 12 | 10,703 | 10,703 | 10,703 | 11,150 | 11,598 | 12,072 | 12,572 | 13,099 | 13,655 | 14,242 | 3,327,872 |
| Increment | 445,788 | 12 | -10,703 | -10,703 | -10,703 | -11,150 | 130,412 | -12,072 | -12,572 | 565,677 | -13,655 | -14,242 | 333,941 |
| **NTFP** | | | | | | | | | | | | | |
| With JFM | 875,531 | 12 | 133,477 | 65,344 | 83,141 | 109,836 | 127,633 | 85,810 | 115,175 | 127,633 | 123,183 | 136,531 | 145,429 |
| Without JFM | 1,252,269 | 12 | 145,474 | 145,429 | 148,658 | 151,887 | 155,325 | 158,972 | 162,841 | 166,948 | 171,306 | 175,930 | 180,838 |
| Increment | -376,738 | 12 | -11,997 | -80,085 | -65,517 | -42,051 | -27,692 | -73,161 | -47,667 | -39,315 | -48,122 | -39,399 | -35,409 |
| **Total** | | | | | | | | | | | | | |
| With JFM | 2,531,635 | 12 | 113,966 | 56,584 | 74,381 | 83,010 | 260,883 | 77,050 | 106,415 | 669,470 | 114,423 | 127,771 | 3,807,242 |
| Without JFM | 2,573,149 | 12 | 156,177 | 156,132 | 159,361 | 163,037 | 166,923 | 171,043 | 175,413 | 180,047 | 184,961 | 190,173 | 3,508,710 |
| Increment | -41,514 | 12 | -42,210 | -99,548 | -84,980 | -80,027 | 93,960 | -93,993 | -68,998 | 489,423 | -70,538 | -62,402 | 298,531 |
| **Internal rate of return** | | | | | | | | | | | | | |
| - to JFM in natural forest | | 8 % | | | | | | | | | | | |
| **Plantation (50 ha)** | | | | | | | | | | | | | |
| Benefits with JFM | 375,501 | 12 | 1,714 | 1,714 | 1,714 | 1,714 | 1,714 | 1,714 | 1,714 | 518,730 | 1,714 | 1,714 | 1,714 |
| Costs | 194,837 | 12 | 64,000 | 154,500 | 6,000 | 6,000 | 6,000 | 6,000 | 0 | 0 | 0 | 0 | 0 |
| Net benefit to JFM | 180,664 | 12 | -62,286 | -152,786 | -4,286 | -4,286 | -4,286 | -4,286 | 1,714 | 518,730 | 1,714 | 1,714 | 1,714 |
| Without JFM | 20,349 | 12 | 3,427 | 3,427 | 3,427 | 3,427 | 3,427 | 3,427 | 3,427 | 3,427 | 3,427 | 3,427 | 3,427 |
| Incremental net benefit | 160,315 | 12 | -65,714 | -156,214 | -7,714 | -7,714 | -7,714 | -7,714 | -1,714 | 515,303 | -1,714 | -1,714 | -1,714 |
| **Internal rate of return** | | | | | | | | | | | | | |
| - to JFM in plantations | | 19 % | | | | | | | | | | | |
| **Total Village** | | | | | | | | | | | | | |
| With JFM | 2,712,298 | 12 | 51,680 | -96,203 | 70,094 | 78,723 | 256,596 | 72,764 | 108,128 | 1,188,200 | 116,137 | 129,484 | 3,808,955 |
| Without JFM | 2,593,498 | 12 | 159,604 | 159,559 | 162,788 | 166,464 | 170,350 | 174,471 | 178,840 | 183,474 | 188,388 | 193,600 | 3,512,137 |
| Increment to JFM | 118,801 | 12 | -107,924 | -255,762 | -92,694 | -87,741 | 86,246 | -101,707 | -70,712 | 1,004,726 | -72,251 | -64,115 | 296,818 |
| **Internal rate of return** | | | | | | | | | | | | | |
| - to JFM in village | | 15 % | | | | | | | | | | | |
| **Revenue Shares** | | | | | | | | | | | | | |
| **Natural forest** | | | | | | | | | | | | | |
| FD | 905,915 | 12 | -19,511 | -8,760 | -8,760 | -26,826 | -8,760 | -8,760 | -8,760 | -36,939 | -8,760 | -8,760 | 2,746,359 |
| FPC | 750,189 | 12 | 0 | 0 | 0 | 0 | 142,010 | 0 | 0 | 578,776 | 0 | 0 | 915,453 |
| **Plantation** | | | | | | | | | | | | | |
| FD | 281,626 | 12 | 1,285 | 1,285 | 1,285 | 1,285 | 1,285 | 1,285 | 1,285 | 389,048 | 1,285 | 1,285 | 1,285 |
| FPC | 93,875 | 12 | 428 | 428 | 428 | 428 | 428 | 428 | 428 | 129,683 | 428 | 428 | 428 |
| **Total** | | | | | | | | | | | | | |
| FD | 1,187,541 | 12 | -18,226 | -7,475 | -7,475 | -25,541 | -7,475 | -7,475 | -7,475 | 352,109 | -7,475 | -7,475 | 2,747,645 |
| FPC | 844,064 | 12 | 428 | 428 | 428 | 428 | 142,438 | 428 | 428 | 708,459 | 428 | 428 | 915,882 |
| **Change** | | | | | | | | | | | | | |
| FD | 812,040 | | | | | | | | | | | | |
| FPC | 844,064 | | | | | | | | | | | | |
| **Employment** | | | | | | | | | | | | | |
| Natural forest | 250 | 12 | 72 | 13 | 13 | 68 | 13 | 13 | 13 | 85 | 13 | 13 | 13 |
| Plantation | 266 | 12 | 80 | 219 | 9 | 9 | 9 | 9 | 0 | 0 | 0 | 0 | 0 |
| Total | 517 | 12 | 152 | 231 | 21 | 77 | 21 | 21 | 13 | 85 | 13 | 13 | 13 |

Note:    Rotations 2 and 3 repeat Years 1-10 of Rotation 1, but it has been assumed that 'without-JFM' degradation continues, as discussed in the text.

# Distributors of World Bank Publications

*Prices and credit terms vary from country to country. Consult your local distributor before placing an order.*

**ARGENTINA**
Oficina del Libro Internacional
Av. Cordoba 1877
1120 Buenos Aires
Tel: (54 1) 815-8354
Fax: (54 1) 815-8156

**AUSTRALIA, FIJI, PAPUA NEW GUINEA,
SOLOMON ISLANDS, VANUATU, AND
WESTERN SAMOA**
D.A. Information Services
648 Whitehorse Road
Mitcham 3132
Victoria
Tel: (61) 3 9210 7777
Fax: (61) 3 9210 7788
E-mail: service@dadirect.com.au
URL: http://www.dadirect.com.au

**AUSTRIA**
Gerold and Co.
Weihburggasse 26
A-1011 Wien
Tel: (43 1) 512-47-31-0
Fax: (43 1) 512-47-31-29
URL: http://www.gerold.co/at.online

**BANGLADESH**
Micro Industries Development
Assistance Society (MIDAS)
House 5, Road 16
Dhanmondi R/Area
Dhaka 1209
Tel: (880 2) 326427
Fax: (880 2) 811188

**BELGIUM**
Jean De Lannoy
Av. du Roi 202
1060 Brussels
Tel: (32 2) 538-5169
Fax: (32 2) 538-0841

**BRAZIL**
Publicações Tecnicas Internacionais Ltda.
Rua Peixoto Gomide, 209
01409 Sao Paulo, SP.
Tel: (55 11) 259-6644
Fax: (55 11) 258-6990
E-mail: postmaster@pti.uol.br
URL: http://www.uol.br

**CANADA**
Renouf Publishing Co. Ltd.
5369 Canotek Road
Ottawa, Ontario K1J 9J3
Tel: (613) 745-2665
Fax: (613) 745-7660
E-mail: order.dept@renoufbooks.com
URL: http://www.renoufbooks.com

**CHINA**
China Financial & Economic
Publishing House
8, Da Fo Si Dong Jie
Beijing
Tel: (86 10) 6333-8257
Fax: (86 10) 6401-7365

**COLOMBIA**
Infoenlace Ltda.
Carrera 6 No. 51-21
Apartado Aereo 34270
Santafé de Bogotá, D.C.
Tel: (57 1) 285-2798
Fax: (57 1) 285-2798

**COTE D'IVOIRE**
Center d'Edition et de Diffusion Africaines
(CEDA)
04 B.P. 541
Abidjan 04
Tel: (225) 24 6510;24 6511
Fax: (225) 25 0567

**CYPRUS**
Center for Applied Research
Cyprus College
6, Diogenes Street, Engomi
P.O. Box 2006
Nicosia
Tel: (357 2) 44-1730
Fax: (357 2) 46-2051

**CZECH REPUBLIC**
National Information Center
prodejna, Konviktska 5
CS – 113 57 Prague 1
Tel: (42 2) 2422-9433
Fax: (42 2) 2422-1484
URL: http://www.nis.cz/

**DENMARK**
SamfundsLitteratur
Rosenoerns Allé 11
DK-1970 Frederiksberg C
Tel: (45 31) 351942
Fax: (45 31) 357822
URL: http://www.sl.cbs.dk

**ECUADOR**
Libri Mundi
Libreria Internacional
P.O. Box 17-01-3029
Juan Leon Mera 851
Quito
Tel: (593 2) 521-606; (593 2) 544-185
Fax: (593 2) 504-209
E-mail: librimu1@librimundi.com.ec
E-mail: librimu2@librimundi.com.ec

**EGYPT, ARAB REPUBLIC OF**
Al Ahram Distribution Agency
Al Galaa Street
Cairo
Tel: (20 2) 578-6083
Fax: (20 2) 578-6833

The Middle East Observer
41, Sherif Street
Cairo
Tel: (20 2) 393-9732
Fax: (20 2) 393-9732

**FINLAND**
Akateeminen Kirjakauppa
P.O. Box 128
FIN-00101 Helsinki
Tel: (358 0) 121 4418
Fax: (358 0) 121-4435
E-mail: akatilaus@stockmann.fi
URL: http://www.akateeminen.com/

**FRANCE**
World Bank Publications
66, avenue d'Iéna
75116 Paris
Tel: (33 1) 40-69-30-56/57
Fax: (33 1) 40-69-30-68

**GERMANY**
UNO-Verlag
Poppelsdorfer Allee 55
53115 Bonn
Tel: (49 228) 949020
Fax: (49 228) 217492
URL: http://www.uno-verlag.de
E-mail: unoverlag@aol.com

**GREECE**
Papasotiriou S.A.
35, Stournara Str.
106 82 Athens
Tel: (30 1) 364-1826
Fax: (30 1) 364-8254

**HAITI**
Culture Diffusion
5, Rue Capois
C.P. 257
Port-au-Prince
Tel: (509) 23 9260
Fax: (509) 23 4858

**HONG KONG, MACAO**
Asia 2000 Ltd.
Sales & Circulation Department
Seabird House, unit 1101-02
22-28 Wyndham Street, Central
Hong Kong
Tel: (852) 2530-1409
Fax: (852) 2526-1107
E-mail: sales@asia2000.com.hk
URL: http://www.asia2000.com.hk

**HUNGARY**
Euro Info Service
Margitszgeti Europa Haz
H-1138 Budapest
Tel: (36 1) 111 6061
Fax: (36 1) 302 5035
E-mail: euroinfo@mail.matav.hu

**INDIA**
Allied Publishers Ltd.
751 Mount Road
Madras - 600 002
Tel: (91 44) 852-3938
Fax: (91 44) 852-0649

**INDONESIA**
Pt. Indira Limited
Jalan Borobudur 20
P.O. Box 181
Jakarta 10320
Tel: (62 21) 390-4290
Fax: (62 21) 390-4289

**IRAN**
Ketab Sara Co. Publishers
Khaled Eslamboli Ave., 6th Street
Delafrooz Alley No. 8
P.O. Box 15745-733
Tehran 15117
Tel: (98 21) 8717819; 8716104
Fax: (98 21) 8712479
E-mail: ketab-sara@neda.net.ir

Kowkab Publishers
P.O. Box 19575-511
Tehran
Tel: (98 21) 258-3723
Fax: (98 21) 258-3723

**IRELAND**
Government Supplies Agency
Oifig an tSoláthair
4-5 Harcourt Road
Dublin 2
Tel: (353 1) 661-3111
Fax: (353 1) 475-2670

**ISRAEL**
Yozmot Literature Ltd.
P.O. Box 56055
3 Yohanan Hasandlar Street
Tel Aviv 61560
Tel: (49 228) 949020
Fax: (977 1) 224 431

R.O.Y. International
PO Box 13056
Tel Aviv 61130
Tel: (972 3) 5461423
Fax: (972 3) 5461442
E-mail: royil@netvision.net.il

Palestinian Authority/Middle East
Index Information Services
P.O.B. 19502 Jerusalem
Tel: (972 2) 6271219
Fax: (972 2) 6271634

**ITALY**
Licosa Commissionaria Sansoni SPA
Via Duca Di Calabria, 1/1
Casella Postale 552
50125 Firenze
Tel: (55) 645-415
Fax: (55) 641-257
E-mail: licosa@ftbcc.it
URL: http://www.ftbcc.it/licosa

**JAMAICA**
Ian Randle Publishers Ltd.
206 Old Hope Road, Kingston 6
Tel: 876-927-2085
Fax: 876-977-0243
E-mail: irpl@colis.com

**JAPAN**
Eastern Book Service
3-13 Hongo 3-chome, Bunkyo-ku
Tokyo 113
Tel: (81 3) 3818-0861
Fax: (81 3) 3818-0864
E-mail: orders@svt-ebs.co.jp
URL: http://www.bekkoame.or.jp/~svt-ebs

**KENYA**
Africa Book Service (E.A.) Ltd.
Quaran House, Mfangano Street
P.O. Box 45245
Nairobi
Tel: (254 2) 223 641
Fax: (254 2) 330 272

**KOREA, REPUBLIC OF**
Daejon Trading Co. Ltd.
P.O. Box 34, Youida, 706 Seoun Bdg
44-6 Youido-Dong, Yeongchengpo-Ku
Seoul
Tel: (82 2) 785-1631/4
Fax: (82 2) 784-0315

**MALAYSIA**
University of Malaya Cooperative
Bookshop, Limited
P.O. Box 1127
Jalan Pantai Baru
59700 Kuala Lumpur
Tel: (60 3) 756-5000
Fax: (60 3) 755-4424
E-mail: umkoop@tm.net.my

**MEXICO**
INFOTEC
Av. San Fernando No. 37
Col. Toriello Guerra
14050 Mexico, D.F.
Tel: (52 5) 624-2800
Fax: (52 5) 624-2822
E-mail: infotec@rtn.net.mx
URL: http://rtn.net.mx

**NEPAL**
Everest Media International Services (P) Ltd.
GPO Box 5443
Kathmandu
Tel: (977 1) 472 152
Fax: (977 1) 224 431

**NETHERLANDS**
De Lindeboom/InOr-Publicaties
P.O. Box 202, 7480 AE Haaksbergen
Tel: (31 53) 574-0004
Fax: (31 53) 572-9296
E-mail: lindeboo@worldonline.nl
URL: http://www.worldonline.nl/~lindeboo

**NEW ZEALAND**
EBSCO NZ Ltd.
Private Mail Bag 99914
New Market
Auckland
Tel: (64 9) 524-8119
Fax: (64 9) 524-8067

**NIGERIA**
University Press Limited
Three Crowns Building Jericho
Private Mail Bag 5095
Ibadan
Tel: (234 22) 41-1356
Fax: (234 22) 41-2056

**NORWAY**
NIC Info A/S
Book Department, Postboks 6512 Etterstad
N-0606 Oslo
Tel: (47 22) 97-4500
Fax: (47 22) 97-4545

**PAKISTAN**
Mirza Book Agency
65, Shahrah-e-Quaid-e-Azam
Lahore 54000
Tel: (92 42) 735 3601
Fax: (92 42) 576 3714

Oxford University Press
5 Bangalore Town
Sharae Faisal
PO Box 13033
Karachi-75350
Tel: (92 21) 446307
Fax: (92 21) 4547640
E-mail: ouppak@TheOffice.net

Pak Book Corporation
Aziz Chambers 21, Queen's Road
Lahore
Tel: (92 42) 636 3222; 636 0885
Fax: (92 42) 636 2328
E-mail: pbc@brain.net.pk

**PERU**
Editorial Desarrollo SA
Apartado 3824, Lima 1
Tel: (51 14) 285380
Fax: (51 14) 286628

**PHILIPPINES**
International Booksource Center Inc.
1127-A Antipolo St, Barangay, Venezuela
Makati City
Tel: (63 2) 896 6501; 6505; 6507
Fax: (63 2) 896 1741

**POLAND**
International Publishing Service
Ul. Piekna 31/37
00-677 Warzawa
Tel: (48 2) 628-6089
Fax: (48 2) 621-7255
E-mail: books%ips@ikp.atm.com.pl
URL: http://www.ipscg.waw.pl/ips/export/

**PORTUGAL**
Livraria Portugal
Apartado 2681, Rua Do Carmo 70-74
1200 Lisbon
Tel: (1) 347-4982
Fax: (1) 347-0264

**ROMANIA**
Compani De Librarii Bucuresti S.A.
Str. Lipscani no. 26, sector 3
Bucharest
Tel: (40 1) 613 9645
Fax: (40 1) 312 4000

**RUSSIAN FEDERATION**
Isdatelstvo <Ves Mir>
9a, Kolpachniy Pereulok
Moscow 101831
Tel: (7 095) 917 87 49
Fax: (7 095) 917 92 59

**SINGAPORE, TAIWAN,
MYANMAR, BRUNEI**
Ashgate Publishing Asia Pacific Pte. Ltd.
41 Kallang Pudding Road #04-03
Golden Wheel Building
Singapore 349316
Tel: (65) 741-5166
Fax: (65) 742-9356
E-mail: ashgate@asianconnect.com

**SLOVENIA**
Gospodarski Vestnik Publishing Group
Dunajska cesta 5
1000 Ljubljana
Tel: (386 61) 133 83 47; 132 12 30
Fax: (386 61) 133 80 30
E-mail: repansekj@gvestnik.si

**SOUTH AFRICA, BOTSWANA**
*For single titles:*
Oxford University Press Southern Africa
Vasco Boulevard, Goodwood
P.O. Box 12119, N1 City 7463
Cape Town
Tel: (27 21) 595 4400
Fax: (27 21) 595 4430
E-mail: oxford@oup.co.za

*For subscription orders:*
International Subscription Service
P.O. Box 41095
Craighall
Johannesburg 2024
Tel: (27 11) 880-1448
Fax: (27 11) 880-6248
E-mail: iss@is.co.za

**SPAIN**
Mundi-Prensa Libros, S.A.
Castello 37
28001 Madrid
Tel: (34 1) 431-3399
Fax: (34 1) 575-3998
E-mail: libreria@mundiprensa.es
URL: http://www.mundiprensa.es/

Mundi-Prensa Barcelona
Consell de Cent, 391
08009 Barcelona
Tel: (34 3) 488-3492
Fax: (34 3) 487-7659
E-mail: barcelona@mundiprensa.es

**SRI LANKA, THE MALDIVES**
Lake House Bookshop
100, Sir Chittampalam Gardiner Mawatha
Colombo 2
Tel: (94 1) 32105
Fax: (94 1) 432104
E-mail: LHL@sri.lanka.net

**SWEDEN**
Wennergren-Williams AB
P. O. Box 1305
S-171 25 Solna
Tel: (46 8) 705-97-50
Fax: (46 8) 27-00-71
E-mail: mail@wwi.se

**SWITZERLAND**
Librairie Payot Service Institutionnel
Côtes-de-Montbenon 30
1002 Lausanne
Tel: (41 21) 341-3229
Fax: (41 21) 341-3235

ADECO Van Diermen EditionsTechniques
Ch. de Lacuez 41
CH1807 Blonay
Tel: (41 21) 943 2673
Fax: (41 21) 943 3605

**THAILAND**
Central Books Distribution
306 Silom Road
Bangkok 10500
Tel: (66 2) 235-5400
Fax: (66 2) 237-8321

**TRINIDAD & TOBAGO
AND THE CARRIBBEAN**
Systematics Studies Ltd.
St. Augustine Shopping Center
Eastern Main Road, St. Augustine
Trinidad & Tobago, West Indies
Tel: (868) 645-8466
Fax: (868) 645-8467
E-mail: tobe@trinidad.net

**UGANDA**
Gustro Ltd
PO Box 9997, Madhvani Building
Plot 16/4 Jinja Rd.
Kampala
Tel: (256 41) 251 467
Fax: (256 41) 251 468
E-mail: gus@swiftuganda.com

**UNITED KINGDOM**
Microinfo Ltd.
P.O. Box 3, Alton, Hampshire GU34 2PG
England
Tel: (44 1420) 86848
Fax: (44 1420) 89889
E-mail: wbank@ukminfo.demon.co.uk
URL: http://www.microinfo.co.uk

**VENEZUELA**
Tecni-Ciencia Libros, S.A.
Centro Cuidad Comercial Tamanco
Nivel C2, Caracas
Tel: (58 2) 959 5547; 5035; 0016
Fax: (58 2) 959 5636

**ZAMBIA**
University Bookshop, University of Zambia
Great East Road Campus
PO. Box 32379
Lusaka
Tel: (260 1) 252 576
Fax: (260 1) 253 952

10/9/97